Victor Soshko
Alexander Soshko

Thermomechanochemical treatment of metals

Victor Soshko
Alexander Soshko

Thermomechanochemical treatment of metals

part 2

ScienciaScripts

Cover image: www.ingimage.com

This book is a translation from the original published under ISBN 978-3-659-52886-6.

Publisher:
Sciencia Scripts
is a trademark of
Dodo Books Indian Ocean Ltd. and OmniScriptum S.R.L publishing group

120 High Road, East Finchley, London, N2 9ED, United Kingdom
Str. Armeneasca 28/1, office 1, Chisinau MD-2012, Republic of Moldova, Europe
Printed at: see last page
ISBN: 978-620-7-90808-0

Table of Contents

Chapter I

Kinetics and mechanism of transformation into hydrogen plasma of surface-active liquid in the cutting zone

1.1 Pyrolytic transformations of the polymeric additive COTS in the cutting zone

It is known that physicochemical transformations of polymer additive occur under the influence of a complex complex of physicochemical phenomena and processes observed in the cutting zone: high temperatures and contact loads, catalytically active surface, emission at the moment of new surface formation of electrons with the intensity of 5000-6000 imp/min, occurrence of thermo-EMF between the tool and the workpiece, etc. All these phenomena can influence the physicochemical transformations of polymer additive. All these phenomena cannot but influence physicochemical transformations of the polymer component of the CRM. Their intensity depends on the type of machining (cutting, drilling, grinding, broaching, etc.), as they mainly create certain kinetic and energy-force conditions of deformation in the contact zone of workpiece-tool-tool-tool-tool. Thus, for example, at stamping the time of contact of the punch with the deformed metal is only 10^{-2} - 10^{-5} s, and the specific pressure reaches 2000-2500 MPa; at turning the average speed of deformation of metal in the cutting zone exceeds by 5-7 orders of magnitude the speed of deformation at static tension or compression and by an order - at shock loading, and the temperature in the deformation zone reaches 0.2 - 0.6 of the melting point of metal.

In order to study the changes in the properties of SOTS in the process of mechanochemical treatment, steel drilling was carried out. At certain time intervals the polymer (PVC), which is the main component of CRM, was sampled and its molecular weight was measured. The obtained data show that the molecular weight (PVC) in the course of the CRM operation is constantly decreasing (Fig.1.1.). These data are indirect evidence of thermodegradation of polymer component of CRM in the cutting zone with formation and accumulation of active fragments of broken macrochains (with lower molecular weight) and low molecular weight products. Studies carried out on different steels (9XC, steel 45, X18H9T) heat-treated for different hardness (HRC 28...32, HRC 42...46, HRC 58...61) have shown that the curve characterizing the function of reduction of molecular weight of polymer, which is a part of COTC, from the depth of drilling, insignificantly changes depending on chemical composition, hardness of processed material and processing tool and practically coincide with the results presented in Fig.1.1.

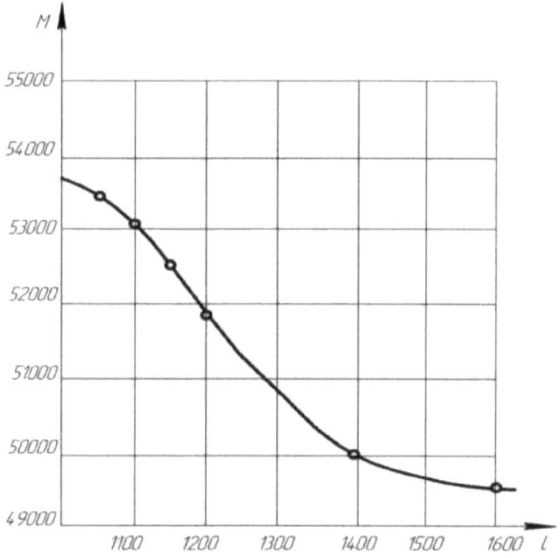

Fig.1.1. Variation of PVC molecular weight in SOTS (M) depending on drilling depth (L) of steel 9XC (HRC 42... .46; n=720 rpm, S= 0.2 mm/rev, drill bit P6M5 0.3 mm).

The products formed as a result of chemical transformations of the polymer macrochain during mechanical processing were analyzed. It should be noted that for PVC as a single polymer such results are known. However, in our case (polymer in the composition of SOTS is in a liquid with several dozens of components dissolved or emulsified in it, high rate of temperature rise in the cutting zone, catalytically active machining surface, etc.) changes in the polymer macrochain may be different. In this connection, the analysis of volatile products resulting from the pyrolysis of PVC and solid residue was carried out under model conditions, when pyrolysis took place in a furnace and the formed chemical elements were in contact with a cell made of different steels.

Experiments have shown that thermal decomposition of PVC during heating in a wide range of temperatures (up to 800°C) regardless of the type of cell material (iron or steel) proceeds in two stages (Fig.1.2).

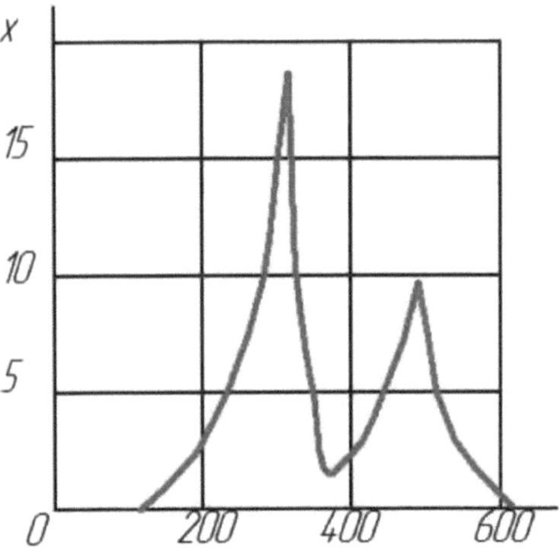

Figure 1.2. Differential yield curve of volatile products($\Delta V/\Delta \tau$) during PVC pyrolysis (heating rate 200° per minute).

This is indicated by the appearance of two distinct maxima at 220-230°C and 450-470°C on the differential yield curve of volatile products. Chromatographic analysis showed (Table 1.1.) that the first stage corresponds to the process of intensive dehydrochlorination of the polymer (almost all hydrogen chloride is detached).

... – CH – CH – CH – CH – ... → HCl

| | | |

H CL H Cl

... – CH = CH – CH – CH – ... → HCl

| |

H Cl

... – CH = CH - CH = CH - CH - CH - ....

| |

H Cl

Table 1.1.

Pyrolysis products of polyvinyl chloride under continuous temperature rise conditions (200°C per min).

Pyrolysis temperature °C	**Pyrolysis products**	**Number of molecules %**
220-230	Hydrogen chloride Benzene Other hydrocarbons	99,5 0,2 0,3
450-470	Ethen Ethane Propane Bhutan Hydrogen Methane Other hydrocarbons	16,6 7,6 4,7 1,6 0,2 0,1 69,2

This reaction is preceded by the detachment of a hydrogen atom

... — CH — CH — CH — CH — CH —... →Ḣ—

| | | | |

H Cl H Cl H

and the formation of a radical

```
... – CḢ – CH – CH – CH – CH -  ...
          |       |      |      |
         Cl      H      Cl     H
```

as well as the detachment of the chlorine atom

```
... – CH – CH – CH – CH – CH - ... Cl̇
        |     |      |      |
       Cl    H      Cl     H
```

and the formation of a radical

```
 ... – CH = CH – CH – CH – CH -  ...
                |      |      |
                H     Cl      H
```

followed by the formation of a conjugated double bond

```
 ... - CH = CH – CH – CH – CH - ...
                |      |      |
                H     Cl     H
```

The second stage (Fig. 1.2.) is caused by the degradation of the formed polymer products containing blocks with conjugated double bonds in macromolecules, and at temperatures above 400°C carbonization occurs, leading to the formation of free carbon [1]. Consequently, during the thermal decomposition of PVC as the polymer decomposition deepens and temperature loads increase, the mechanism of the gross process becomes much more complicated. PVC decomposition is observed already at low temperatures and under conditions of continuous temperature rise, the dehydrochlorination process is completed by about 250°C. Above 400°C the polymer degradation proceeds with pre-dehydrochlorinated polymer residue and corresponds to pyrolysis of hydrocarbons, which leads to formation of a mixture of highly active gaseous products of saturated and supersaturated hydrocarbons, hydrogen, liquid hydrocarbons (Table 1.1.) and carbonaceous residue with high carbon content. It should be noted that in the process of machining with the use of polymer-containing coolants, carbon residue always accumulates on the cutting edges of the tool.

There are reasons to believe [2] that the mechanism and kinetics of thermal degradation of the polymeric component of the model composition of the CRM largely corresponds (at least qualitatively) to the mechanism of chemical transformations of the polymeric component of CRM under conditions of mechanical processing of metals. This means that when polymer-based CRM enters the cutting zone under the action of high temperatures, the polymer component is destructed with the formation of gaseous, liquid and solid products, which cause a significant increase in the efficiency of CRM as compared to CRM with only low-molecular components.

When comparing the difference in the behavior in the zone of mechanical treatment of high-molecular compounds and low-molecular hydrocarbons, which are the basis of commercial coolants, at least two following features can be distinguished. Firstly, low-molecular hydrocarbons, which are usually included in the composition of commercial coolants, practically do not undergo decomposition up to a temperature of about 190°C [2-4] in an unclosed volume and normal pressure. Therefore, pyrolysis of low molecular weight components of CRMs cannot occur in the zone of machining, and hence there are no active chemical elements in the zone of machining that affect the machinability of metal.

However, a definitive answer to this question can only be given when studies are conducted directly in the treatment zone. We have conducted such studies in a specially made hermetic chamber and the results are discussed in the following sections. Here it should be noted that, taking into account the extreme nature of the process, the presence of active products in the treatment zone can be assumed, but their concentration will be extremely insignificant [5].

Assuming this possibility, we note that such reactions for polymers, in contrast to low-molecular-weight hydrocarbons, follow a chain mechanism[5, 6]. Therefore, the rate of formation and accumulation of active products in the cutting zone for the first case is much higher than for the second case.

Thus, the main difference between the behavior in the zone of mechanical treatment of SOTS based on high-molecular compounds and SOTS based on low-molecular media is that in the first case at much lower temperatures the destruction of the polymer macrochain occurs with the release of hydrogen in the active form and other various gaseous media of high activity and concentration. The reaction of thermomechanodegradation of polymer additive COTC occurs with high speed and has an "explosive" character, while in the case of low-molecular hydrocarbons their decomposition, if it occurs, then at much higher temperatures. Probably, it is this difference that causes a significant change in the mechanism of influence of polymer-based COTS compared to conventional compositions.

It should be noted that the chain mechanism of depolymerization of the macrochain of PVC polymer at relatively low temperatures is also inherent in other polymers. Thus, for example, when turning steel and a model medium based on polymethyl methacrylate (PMMA), a continuous decrease in the molecular weight of the polymer is observed (Fig. 1.3.). Moreover, the intensity of molecular weight decrease is the greatest at the beginning of the experiment, which characterizes the high sensitivity of the polymer component to temperature change. Similar results were obtained when measuring the relative integral intensity of NMR signals of the polymethylmethacrylate group in the process of turning (Fig. 1.4), these data suggest that various high molecular weight compounds can be used as additives in SOTS.

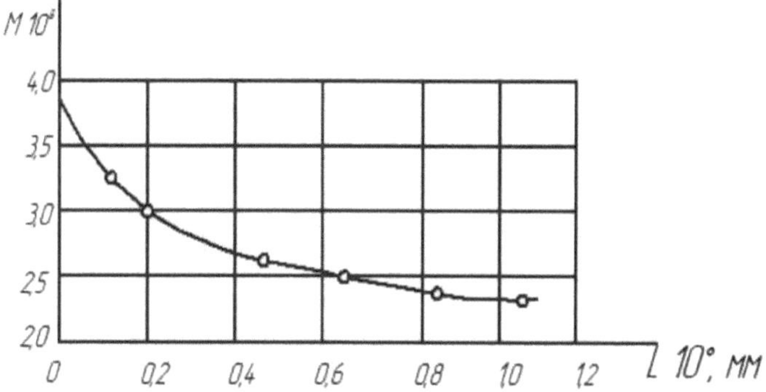

Fig. 1.3: Variation of PMMA molecular weight (M) as a function of turning length (L) of 9HS steel (HRC 42.46; cutting mode: V = 2.2 m/s, S = 1 mm/rev, t = 1.5 mm, T15K6 cutter).

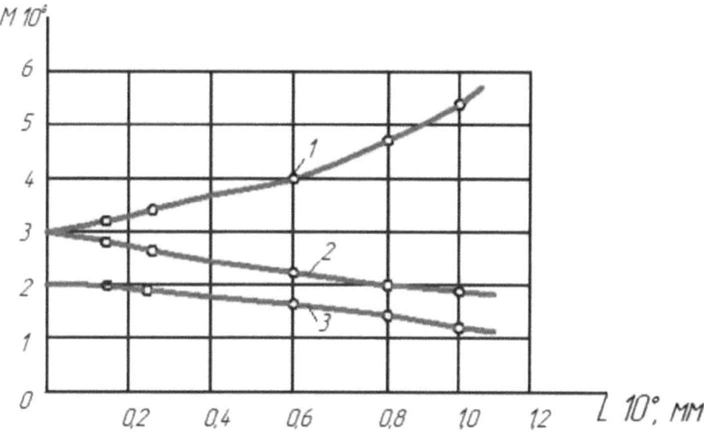

Fig. 1.4. Change of relative integral intensity of NMR signals (Δ) of PMMA groups during turning of steel 9XC, 1 - groups - CH3; 2 - groups COOCH3; 3 - groups CH2 (Cutting mode is the same as in Fig. 1.6).

1.2. Composition of chemical elements formed in the cutting zone

Within the framework of this work, the research was concentrated mainly on a detailed study of the composition of gaseous pyrolysis products of polymer additives and model compounds in order to reveal the influence of these or those components of gas mixtures on the process of mechanical treatment of metal. The first experiments carried out in this direction showed that some gases released during pyrolysis of polymeric additives (emulsions of polyethylene and polyvinyl chloride) demonstrate tribological activity. This activity is manifested in the reduction of cutting force, torque, increase in the wear resistance of cutting tools, as well as in the reduction of energy consumption for OMD [7].

Since the chain of pyrolytic transformations of the initial additive leads to the formation of carbon and hydrogen in atomic or other active forms (radicals, ions, ion-radicals), the hypothesis [7] of permanent carbonization of the cutting edge of the tool, hydrogenation of the workpiece material and chips, and active participation of hydrogen in the mechano-chemical process during metalworking in the medium of coolants was put forward. This conclusion is of great practical importance, since from here follows direct recommendations for the search of effective additives of STC among polymeric compounds that give active forms of carbon and hydrogen in the chain of thermomechanical transformations.

In spite of the fact that already at the first stage of the work versatile and very encouraging experimental data were obtained in favor of this hypothesis, it certainly needed further elaboration and testing.

At the same time, it is possible to identify a number of problems, the solution of which will make it possible to move in this issue in the right direction. Among them, the most important is to find out the detailed composition of gaseous products of polymer additive transformation in the cutting zone and evaluation of tribological activity of these gases.

The gaseous products formed during pyrolysis and mechanochemical degradation of polyethylene emulsion OXALEN-ZO, polyethylene (PE) of molecular weight 100000 produced on vanadium catalyst, polyvinyl chloride emulsion and polyvinyl chloride were investigated in this work.

Pyrolysis was carried out in quartz ampoules. A 0.5g polymer suspension was placed at the bottom of the ampoule and vacuumized with slight heating (50°C). The evacuated ampoule was placed in a muffle furnace preheated to 800-850°C and kept until complete charring of the sample (1-2 min.). The pyrolysis products formed during this process were condensed in the cold part of the ampoule. After pyrolysis, the ampoule was filled with helium to atmospheric pressure to further introduce its contents into the syringe chamber of the chromatograph.

The gas phase formed during drilling of steel in the medium of the corresponding SOTS was also analyzed. The gases under study were sampled in a standardized manner using a calibration streamer into a volume (1L) within 1.5 min. All volatile products were first collected at liquid nitrogen temperature (-196°C) into a one-liter volume. The condensed fraction (condensation time 30 min) was concentrated by refreezing it into a small volume (10 cm^3) followed by filling the ampoules with helium for chromatographic analysis. The gaseous fraction that does not condense at liquid nitrogen temperature was not analyzed.

Drilling of 1X18Ni9T steel in COTS medium was performed with a P18 drill d=5 mm at a drill rotation speed of 1500 rpm. A thin layer of PE emulsion was applied to the bottom of the steel cup and covered with a lid, which had two holes: for drill entry and extraction of gaseous products generated during steel drilling.

The gas phase was analyzed on the gas chromatograph CHROM - 5 (70°C, Rogarak-R, carrier gas flow rate 30 ml/min.), the high sensitivity of the chromatograph allows to identify gases in the amount of at least 10 mol. Table 1.2 shows the results of chromatographic analysis of gaseous products of thermal and mechanochemical decomposition of PE emulsion. Each figure corresponds to the average value of three measurements.

Table 1.2.

Chromatographic analysis of gaseous products released in the processes of PE emulsion pyrolysis and steel drilling in COTS medium.

№	Pyrolysis		Drilling		Retention times of individual substances sec.	
	PE wm	PE	PE em	PE+water		
1	17	16	16	17	17	methane
2	-	29	28	-	30	ethylene
3	30	33	33	34	34	ethane

4	99	100	96	105	96	propylene
5	145	-	-	-	105	propane
6	263	260	257	264	-	-
7	400	400	400	410	400	isobute n
8	575	-	-	-	-	
9	1260	-	1280	1200	1290	acetone
10	1650	1620		-	1620	pentene
11	4380	4350	-	-	4350	hexene
12	-	6690	-	-	6630	hexane

Table 1.2. shows that the composition of gas mixtures formed during pyrolysis and drilling in the area of volatile fractions is more or less identical, while the high-boiling fractions differ significantly. Fractions 10, 11, 12 are absent in the drilling products, which can be explained by their active decomposition or/and absorption on the juvenile surface of steel.

Similar data were obtained in experiments on pyrolysis and drilling in the medium of polyvinylchloride emulsions. The only difference is the presence of chlorinated hydrocarbons in the gas phases.

1.3. Some regularities of high efficiency of polymeric additives for COTCs

The above studies have shown that in contrast to traditional low-molecular-weight coolants the optimum modes of application of polymer-containing media for metal cutting and pressure treatment are in the area of higher temperatures.

This is caused by the fact that with increasing temperature the rate and depth of the degradation process of high-molecular-weight additives in SOTS increases, and hence the yield of low-molecular-weight radical active substances that have a direct effect on machining increases. In this connection, finding the temperature zones that correspond to the optimum concentration of the yield of low-molecular-weight products will make it possible to determine the optimum processing modes.

The processes of radical and polyconjugated systems formation during polymer pyrolysis and their physicochemical characteristics were studied by electron paramagnetic resonance (EPR), as well as by measuring dielectric and electrophysical characteristics. To study the mechanism of hydrogen formation from polymer molecules in the process of their use in metalworking, the method of nuclear magnetic resonance (NMR) was used, and for similar processes occurring during thermal degradation of polymers was carried out in the current of inert gas (argon), at a rate of up to 40 ml/min and in air, in a quartz cell at temperatures of 300-900 °C. The mass of the sample was 20 mg.

The solid residue formed during pyrolysis was analyzed on an EPA-2A EPR spectrometer. For this purpose, the product sample was placed in a quartz ampoule in the resonator of the spectrometer. Diphenylpicrylhydrazyl (DPPH) was used as a reference for the EPR signal intensity, and MgO was used as a reference for the magnetic field strength.

A special cell was used to study the dielectric and electrophysical properties of the solid residue. Punches, between which the sample was placed, were electrodes and compressed the sample to a pressure of 395 kG/mm".

The pressure value was chosen in order to eliminate the measurement error in the properties of the powdered pyropolymer residue, which can be caused by the looseness between the particles. Such a pressure value corresponds to the conditions of conductivity saturation, i.e. the conditions under which further pressure increase does not change the electrical conductivity of the sample tablet. Measurements of the properties of the pyrolyzed polymer were carried out in the temperature-time range of 25-250°, every 5°, with holding at each temperature. Determination of the percentage of free radicals in the total number of paramagnetic centers of the material was carried out according to the following procedure.

A 0.02% weight solution of DFPG in benzene was prepared and the EPR signal intensity of 0.1 mL DFPG solution was determined. A suspension of pyrolyzed polymer was placed in a container with DFPG solution, then removed after 20 min and the EPR signal intensity of the given volume of DFPG solution was measured again. By the deviation of the difference in the EPR signal intensities of the DFPG solution before and after the pyrolyzed polymer suspension was added to it, the percentage of free radicals in the total number of paramagnetic centers was determined.

Nuclear magnetic resonance (NMR) spectra were studied on a Tesla B8-487c radio spectrometer. The operating frequency of the spectrometer is 80 MHz. Tetramethylsilane (TMS) was used as a reference. The studies were carried out at room temperature.

The materials used for the research were polystyrene (PS), polyvinyl chloride (PVC), low pressure polyethylene (PE) and solvent: benzene - CH, carbon tetrachloride - CH.

The EPR signal of pyrolyzed polymers was typically a 98-10Gs wide singlet with a g-factor close to that of a free electron (Fig. 1.5.(a)).

Fig. 1.5. View of the EPR spectrum of polymer heat treatment products

We obtained a similar singlet in the studies of PS (Fig. 1.5.(a)), PE (Fig. 1.5.(b)), and PVC (Fig. 1.5.(c)), and the width of the singlet and its position are close for all polymers studied.

The EPR singlet of this kind is also attributed to stable radicals that can be retained in the solid matrix of the pyrolyzed polymer in positions in which there are steric hindrances to their recombination. This signal is attributed to the presence of a conjugated system of the form.

$$- CH \quad CH - CH \quad CH - C - CH \quad CH - CH \quad CH -$$

The absence of superfine structure (STO) in this case is due to the retardation of radical rotation in a rigid matrix. The EPR signal of this type is often observed in studies of pyrolyzed polymers forming a polyconjugated polycyclic system consisting of graphitized aromatic rings.

The kinetics of formation of polyconjugated systems is interesting from the point of view of establishing optimal modes of mechanochemical treatment of solids. This is evidenced at least by the fact that the formation of polyconjugated systems in the pyrolysis of polymer, there is a continuous release of hydrogen, which has a significant impact on the process of destruction of metal in the treatment zone.

One of a number of going simultaneously with the reaction of formation of polycyclic system at thermodestruction of PE can be represented in the form:

$$CH{=}CH{-}CH_2{-}CH_2 \rightarrow CH_2{-}CH(C_5H_{10})$$

$$CH_2{-}CH_2{-}CH(C_5H_{10}) + CH_2{=}CH{-}CH{=}CH_2 \rightarrow CH_2{-}CH_2{-}C(\ldots)CH_2 + nH_2\uparrow$$

The ultrafine structure of the PE signal can be compared to a sextet of lines with rather poorly resolved SOTS. Such a structure of the spectrum has a radical of the form

$– CH_2 – CH – CH_2 –$

At the same time, PE end radicals formed during mechanodegradation have such a spectrum.

For polystyrene, the appearance of the EPR signal can correspond to radicals

$CH_2{-}\dot{C}(C_6H_5){-}CH_2$ или $CH_2{-}C(CH_3)(C_6H_5)\bullet$

To elucidate the radical species formed, studies were carried out by nuclear magnetic resonance (Fig. 1.6.) of products formed as a result of drilling of steel 45 in a solution of PS in CCl_4 . It was found that as a result of thermomechanodegradation of the polymer-containing medium in the metal processing zone, the proton balance in the polystyrene chains changes significantly, there is a sharp increase in the number of protons of benzene rings in the chains in relation to the protons of the main chain. This fact corresponds to the above provided mechanism of hydrogen atoms detachment exactly from the main chain of polystyrene macromolecules.

Such dependence of the kinetics of paramagnetic centers appearance on temperature and time parameters of heat treatment was investigated.

It was shown that the dependence of the concentration of paramagnetic centers on the time of heat treatment both in air and inert gas has an extreme character (Fig.1,7). Moreover, with the increase of the treatment temperature, the maximum concentration of paramagnetic centers narrows and shifts to the region of smaller treatment times, i.e., a certain temperature-time analogy of this process is observed. After a certain period of time or at increasing temperature, the process of opening of double bonds begins to prevail in the pyropolymer, which causes a decrease in the concentration of PMCs.

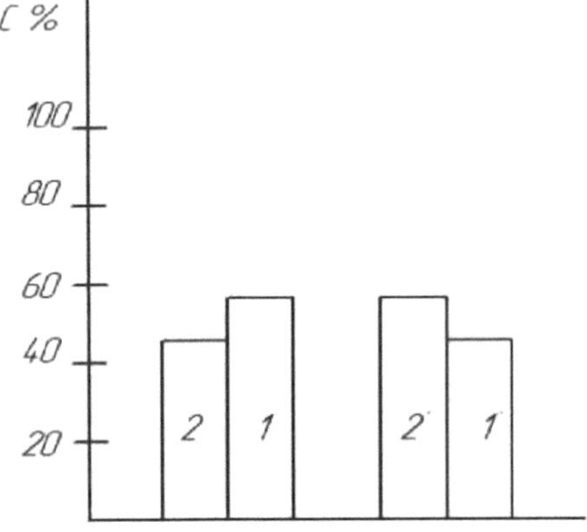

Fig. 1.6. NMR signal intensity of PS solution in *CCl* 1 - signal corresponding to protons of the polymer main chain, 2 - signal corresponding to protons of the polymer aromatic ring; signals 1, 2 - before treatment, 1', 2' - after treatment.

To assess the possibility of the influence of polymer decomposition products on the treatment process, it was interesting to compare the kinetics of the yield of volatile products during the thermal destruction of polymers, and the formation of paramagnetic centers in the polymer and stable free radicals in this system. From Fig. 1.7, it can be seen that in the case of polystyrene, in the initial period of time, there is an output of volatile products, then formed paramagnetic centers having the character of free radicals, and after that, in the next period of time, the character of paramagnetic centers changes, among them appear PMCs are not stable radicals. Apparently, with the further course of thermal exposure time, polysconjugated systems are formed. Eventually, double bonds are opened in the material and a cross-linked system without paramagnetic activity is formed.

It is characteristic that a certain interval between the time of formation of volatile products and the time of occurrence of PMC in the pyrolysis of PS is maintained at higher temperatures. For PE, the peak of total paramagnetic activity coincides to some extent with the maximum of free radical concentration. This makes it possible to assume that the formation of radical systems stabilized by conjugation of the kind shown in Fig. 1.5(6) is probable during PE heat treatment.

Since the heat treatment of polystyrene was carried out in air, it was natural to assume that the responsible for the death of free stable radicals in PS was air oxygen interacting with free radicals.

R%, вес

$tg_{max}\delta \cdot 10^4$ $C_{пмц}$, отн.ед. C, % летучих продуктов

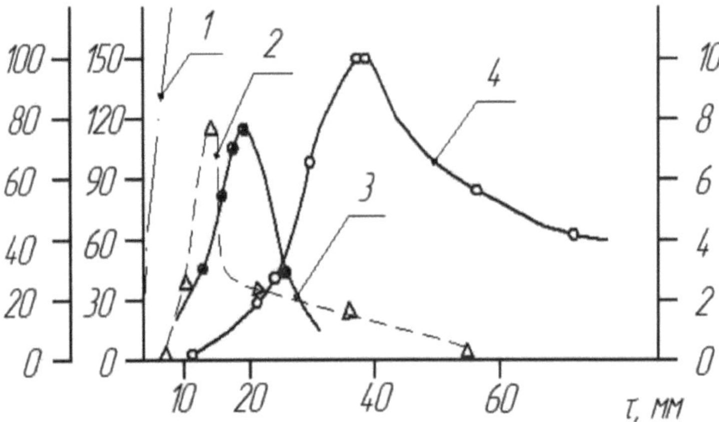

Fig. 1.7. Concentration dependence of products formed during thermo-oxidative degradation of polystyrene: 1-styrene, 2-proportion of free radicals in the pyroostat matrix in relation to the total number of PMCs, 3-groups C=0 (determined by the dependence of tg S on τ), 4-paramagnetic centers of the type of polysconjugated systems.

To verify this assumption, dielectric studies of the temperature dependence of *τga of* the pyrolyzed polymer on the measurement temperature were carried out.

At the formation of C=0 bonds in polymers at measurements of dielectric properties at the frequency of 1000 Hz and at the measurement temperature of 33±3°C the maximum *tdb* is registered, the value of which is directly proportional to the concentration of C=0 bonds. In this connection the dependences *τgs* = *f(T°}* for pyrolyzed PS were investigated. According to these dependences the graph of intensity of maximum *τg6* at T=33°C depending on the time of heat treatment of PS in air was plotted.

The graph in Fig. 1.8 shows that during the time during which the death of free stable radicals occurs, a significant number of oxygen-containing groups of the C=0 type are formed in the system. When passing through a certain value of the time parameter, the number of these groups decreases and simultaneously the total concentration of polyconjugated systems in the sample increases.

In order to check whether oxygen addition is a decisive factor in the observed character of the kinetic dependences of the formation of the total number of PMCs, similar studies were carried out with heat treatment of the polymer in an inert gas.

It is shown that the qualitative character of mutual location of maxima of volatile yields and total PMC concentration remains the same, i.e. some interval between the peaks of these dependences on processing time is preserved.

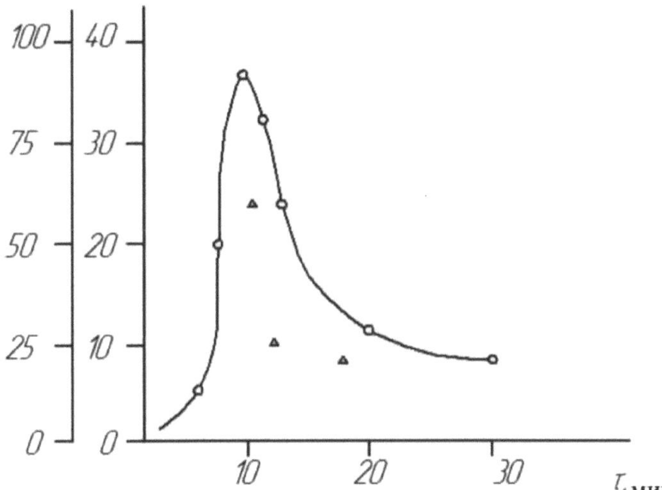

Fig. 1.8. Concentration dependences of polyethylene thermo-oxidative degradation products on treatment time: 1 - free radicals, 2 - paramagnetic centers of the type of polyconjugated systems.

Consequently, it can be assumed that the formation of oxygen-containing groups is not exclusively responsible for the death of free stable radicals in the pyrolyzed PS system. However, during the period from the beginning of stable radical formation to the appearance of the maximum concentration of total PMCs, pyrolyzed polystyrene actively interacts with air oxygen. Obviously, this also produces a significant amount of molecular substances of peroxide character, which affect the metalworking process by chemisorbing on the metal and facilitating the processes of formation of new solid surfaces in the process of destruction.

It is known that there is a close relationship between magnetic, electrophysical and catalytic properties of organic polymer products. For example, heat-treated polyacrylonitrile catalyzes the dehydrogenation and decomposition reactions of various organic substances (fats, formic acid, etc.). At the same time, the catalytic properties of polymeric semiconductors unambiguously correspond to their electrophysical properties. Taking this into account, studies of electrophysical properties of solid products of polymer heat treatment were carried out.

Fig. 1.9. shows the dependences of the logarithm of the electrical resistance of pyrolyzed polymer on the inverse temperature of measurements, for different heat treatment regimes.

It is seen that for all investigated polymers with increase in temperature of processing there is a decrease in electrical resistance of the heat-treated product (residue). The presence of two sections of the dependence $lg\,p = f(1/T°)$ *is* determined, apparently, by the fact that the initial section of the dependence is caused by impurity conductivity, and the next by the intrinsic conductivity of the polymer semiconductor. Based on the value of the slope of the dependences $Ig\ p = f(1/T°)$ for the region of intrinsic conductivity, we determined the width of the forbidden conduction zone of polymer pyrolysis.

The size of the forbidden zone of the electrical conductivity of semiconductors is an important criterion of their catalytic properties. Depending on the type of catalyzed reactions, either the reaction is accelerated or slowed down as ΔE decreases.

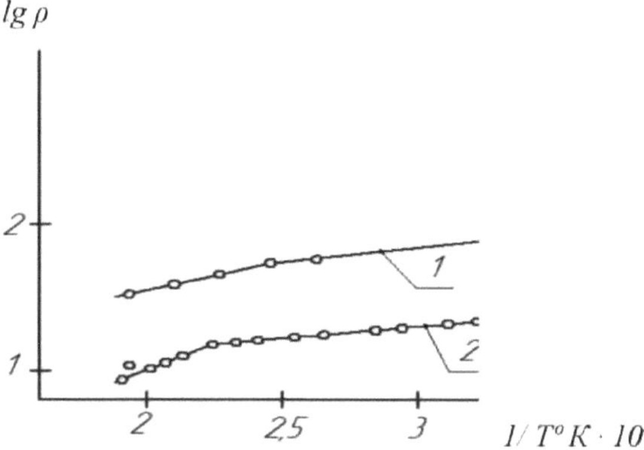

Fig.1.9. Dependence of electrical resistance of PVC thermal destruction products on temperature in helium atmosphere during 1 min, 1 - $T°$ = 600°C; 2 - $T°$ = 700°C.

In order for the reaction to accelerate with decreasing ΔE, it is necessary that it be of the donor type (i.e., accelerate as the Fermi level of the catalyst decreases toward the forbidden zone). Since a significant part of chemical reactions of polymer decomposition can be attributed to donor-type reactions, it was naturally assumed that the dependence of ΔE on pyrolysis time at different temperatures would be indicative in elucidating the temperature-time modes of maximum catalytic activity of polymer pyrolyzate. Fig. 1.10. shows such dependences on heat treatment time. It can be observed that the maximum of the forbidden zone width of the electrical conductivity and the maximum of the total PMC concentration coincide in the same time domain. The fact that two characteristics of the substance, electrophysical and magnetic, which are both related to the catalytic activity of this substance, coincide in the same region for the polymer pyrolyzate suggests that this region is the region of maximum catalytic activity of the polymer during its mechanochemical treatment. In this case, the polymer catalyst formed participates in catalyzing various reactions of thermodestruction of macromolecule fragments, eventually leading to autoacceleration of these reactions, resulting in hydrogen release, which has a significant effect on the destruction and deformation of metals in the process of their mechanical treatment. In the presence of oxygen, the pyropolymer residue creates optimal conditions for the formation of peroxide groups, which also influence the process of mechanical treatment of solids.

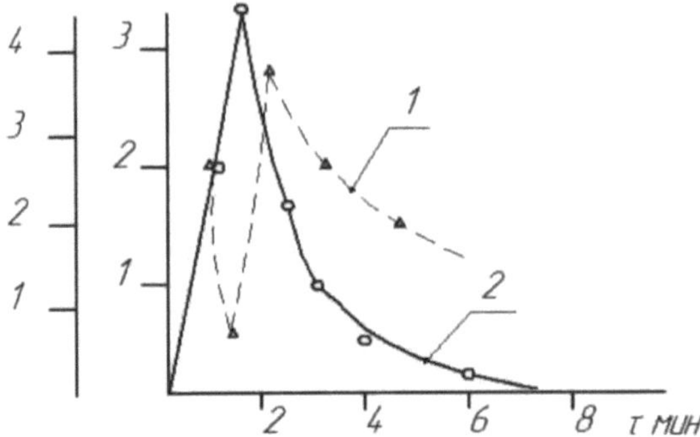

Fig. 1.10. Dependence of concentration of paramagnetic centers and activation energy of electrical resistance of PMMA thermal destruction product on the time of heat treatment at T =600°° C 1-width of the forbidden zone of electrical conductivity, 2-concentration of PMCs

In this regard, in order to establish the optimal modes of using polymers as technological media for mechanical processing of metals, the temperature and time regions of maximum catalytic activity of polymers were revealed. This area is in the interval of small values of time during which there is an optimal concentration of degradation products (up to 100 sec) and relatively high temperatures (500-600°C). (Fig.1.11.). In the same region lies the minimum ΔE of conductivity of these substances.

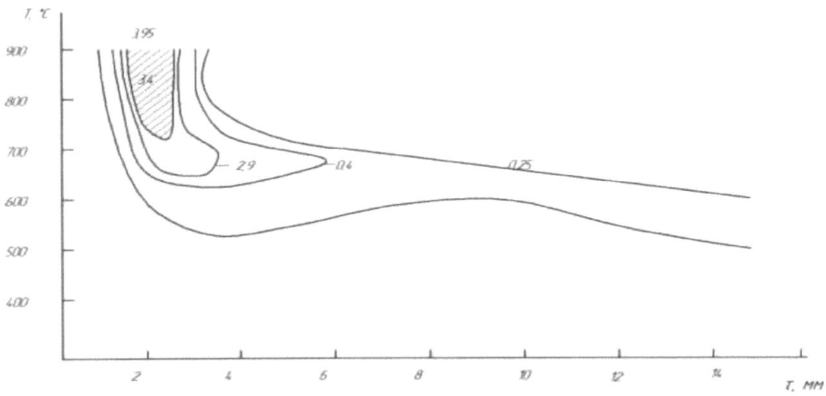

Fig. 1.11. Zones of the same PMC concentration during thermal destruction of PS in Ne atmosphere.

To a certain extent, the conducted studies allow to explain the fact that polymer-containing media work most effectively in severe modes of metal processing, when significant temperatures occur in the center of deformation and fracture, and the treated material is in the zone of mechanical action for a minimum time.

Chapter II

Basic information about the hydrogen that causes the effect to occur

Numerous research results and their analysis presented in the previous chapters, as well as the experience of application of polymer-containing coolants in industry testify to the fact that high efficiency of such coolants is due to the influence of hydrogen on chip formation processes. It has also been established that the active form of hydrogen is formed in the cutting zone as a result of successive pyrolytic transformations of the polymeric component of CRM under the effect of temperature in the machining zone, which arises as a result of friction of the tool with the workpiece and plastic deformation in the cut metal layer.

Due to the fact that the chemical element hydrogen is crucial in the realization of the mechanochemical effect, and information about it is presented only in special literature and students' ideas about it are limited to the course of general chemistry, so this chapter gives the basic physical properties of hydrogen.

It should be noted that despite the variety of forms and types of manifestation of hydrogen influence on the deformation and fracture processes of steel and alloys in general, this influence is determined by the hydrogen content, the nature of the interaction of metals and alloys with hydrogen, the state of hydrogen in the metal, the magnitude of the acting stresses (external or internal), the scheme of the stressed state. Hydrogen can also influence crack initiation, crack propagation or both of these fracture stages. Such a variety of factors leads to the fact that there is probably no single mechanism of hydrogen embrittlement of metals, and even for one metal, the current mechanism of hydrogen influence can change when changing the above factors. The possibility of simultaneous action of several mechanisms is also not excluded.

When analyzing the influence of hydrogen on deformation and fracture processes, it is necessary to distinguish between primary and secondary factors. The primary factors include hydrogen sources, the processes of hydrogen atoms transportation from the source to some region (in which certain hydrogen embrittlement mechanisms can be realized), hydride formation, lattice decohesion, interaction of hydrogen atoms with dislocations, and formation of non-plosions. The second factors include various impurities, surface hydrides and oxide films, plastic blunting of the crack, and the degree of triaxial stresses.

Based on the studies devoted to various forms and types of manifestation of hydrogen influence on the physical and mechanical properties of steels and especially to the processes and phenomena occurring during their deformation and fracture (critically analyzed in monographs [8-13] and review articles [14-16]), four groups of hypotheses can be distinguished:

1) high-pressure theories of molecular hydrogen including the theory of maximum triaxial stresses [10];

2) adsorption theory, according to which hydrogen facilitates the nucleation and propagation of cracks due to a decrease in surface energy [8,9];

3) theory of reduction of cohesive strength of metals by dissolved hydrogen [11,13];

4) theories based on the assumption that hydrogen affects not only the fracture processes but also the mechanism of the preceding plastic deformation [14,15,17].

At the same time, the proposed hypotheses cannot explain all the regularities of hydrogen influence, in particular, difficulties arise, for example, when explaining reversible hydrogen embrittlement. In essence, these hypotheses explain the effect of hydrogen pre-dissolved in steel on the deformation and fracture processes. While to explain the mechanism of hydrogen influence formed as a result of pyrolytic transformations of the polymeric component of SOTS directly in the cutting process it is necessary to take into account: chemical interaction of hydrogen with the external surface of the metal before its deformation, with the internal surfaces of grains, blocks, non-metallic inclusions during deformation, the ratio of deformation rate and diffusion rate of hydrogen, its interaction with dislocations, etc., as well as the chemical interaction of hydrogen with the external surface of the metal before deformation, with the internal surfaces of grains, blocks, non-metallic inclusions during deformation.

In this connection, it seems reasonable to identify the general basic regularities of hydrogen influence, characteristic both during deformation and fracture of steel and for the case of preliminary electrolytic hydrogenation with subsequent deformation of steel, regardless of whether hydrides are formed in the metal or not, whether the cohesion of the lattice decreases or increases during hydrogen dissolution, etc. There must be something common in the mechanism of hydrogen influence, which does not depend on their crystal structure, nor on the internal pressure of hydrogen, nor on the presence of hydrides.

In our opinion, common in the processes leading to changes in metal properties under the influence of hydrogen may be: a) dislocation nature of the mechanism of plastic deformation; b) directed diffusion of hydrogen into the zones of triaxial stretching.

Proceeding from the above, for the interpretation of the mechanism of hydrogen influence on the process of chip formation during steel machining we used modern ideas about the influence of electrolytic hydrogenation of steel, both before and during its deformation, as well as the results of our own research, the main of which are presented in the monograph [7], as well as the data obtained together with our collaborators and published in [15, 17].

Hydrogen is a very widespread element. It is a constituent of most of the substances that make up living organisms, as well as many inorganic substances. More compounds of hydrogen are known than of any other element; carbon ranks second in the number of compounds, only slightly behind hydrogen.

In its free state, hydrogen H2 is a gas that has no color, odor, or taste. It is the lightest of all gases, its density is about $/^{1}{}_{4}$ the density of air. Its melting (- 259°C) and boiling (- 252.7°C) temperatures are very low; only helium has even lower temperatures. Liquid hydrogen, with a density of 0.070 g - cm^{-3} , is, as might be expected, the lightest liquid. Crystalline hydrogen, which has a density of 0.088 g\cm^{-3} , is also the lightest crystalline substance. Hydrogen is very poorly soluble in water: in 1 liter of water at 0 ° C and a pressure of 1 atm dissolves only 21.5 gaseous hydrogen. Solubility decreases with increasing temperature and increases with increasing gas pressure.

2.1. Electronic structure of hydrogen

The smallest and lightest nucleus is the proton. The proton carries a single positive charge and together with one electron, which carries a single negative charge, forms a hydrogen atom.

Soon after the ideas of the presence of a nucleus in each atom were developed, ideas were advanced as to exactly how the proton and electron combine to form the hydrogen atom. The mutual attraction of the oppositely charged electron and proton could be expected to result in the electron orbiting around the much heavier proton, much like the Earth orbits around the Sun. Bohr assumed that the orbit of the electron in an ordinary hydrogen atom should be circular with a radius of 53 pm. According to the calculation, the electron should orbit this orbit at a constant speed of 2.18 - 106 $M{\cdot}C^{-1}$, which is slightly less than 1% of the speed of light.

The results of the investigations of many physicists show that this picture is correct only approximately. The electron does not move along some definite orbit, but performs to a certain extent disordered motion - sometimes it appears very close to the nucleus, sometimes considerably distant from it. Moreover, it moves mainly toward or away from the nucleus and moves in all directions relative to the nucleus rather than being in one plane. Although it does not remain exactly 53 pm from the nucleus, this distance still determines its most probable position relative to the nucleus. Due to its rapid motion around the nucleus, it effectively occupies all the space within a radius of about 100 pm from the nucleus and thus predetermines the value of the effective radius of the hydrogen atom to be about 100 pm. It is this electron motion that makes atoms, consisting of particles with diameters of only ~10^{-3} pm, behave as solid objects with diameters of several hundred pm. The velocity of an electron in a hydrogen atom is not constant; its Bohr mean value is $2.18{\cdot}10^{6}$ $M{\cdot}C^{-1}$.

Thus, it can be stated that a free hydrogen atom has a heavy nucleus in the center of a sphere bounded by a space filled by an electron moving rapidly around the nucleus. The diameter of such a sphere is about 200 pm.

Based on the equations of quantum mechanics describing the electron of the hydrogen atom in the normal state, it was concluded that it is incorrect to say that the electron moves around the nucleus along some orbit. Instead, it is customary to say that the electron occupies an *orbital. The* orbital occupied by an electron of a hydrogen atom in the normal state (the most stable state) is called the ls-orbital. The number 1-in this case, the value of the *principal quantum number p.*

There is only one orbital for n=1. In the case of the hydrogen atom, there are other possible orbitals corresponding to n=2, n=3, etc. A hydrogen atom in which an electron occupies one of these orbitals is unstable; it is assumed that such an atom is in an *excited state.* To move a hydrogen atom from the normal state to the first excited state (n=2), it is necessary to expend a large amount of energy equal to 75% of the energy required for complete detachment of an electron. The diameter of the atom in this excited state is about 800 pm, i.e., it is four times the diameter of the atom in the normal state.

2.2. Hydrogen molecule

The simplest example of a covalent molecule is the hydrogen molecule H_2 . The electronic structure of the molecule is written as H : H; this writing indicates that two electrons simultaneously belong to both atoms, forming a bond between them. This structure corresponds to the structure with an H-H valence bond.

The hydrogen molecule, as it was established as a result of studying its spectrum and calculations based on the theory of quantum mechanics, has the structure shown in Fig. 2.1. Two nuclei in the hydrogen molecule are rather firmly held at a distance of about 74 pm; at room temperature, the amplitude of deviation from the average distance between them varies within a few picometers and slightly increases at elevated temperatures.

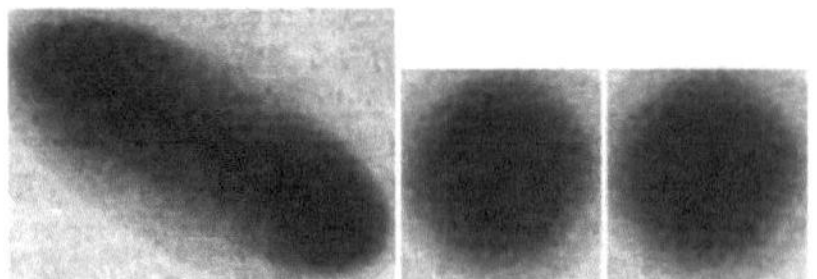

Figure 2.1. Distribution of electrons in a hydrogen molecule and two hydrogen atoms.

The distance between the two nuclei in this molecule is 74 pm.

The two electrons move very fast in a certain region surrounding the two nuclei; the average distribution of electrons in time is shown by the shaded areas in Fig. 2.1. It is easy to see that the motion of the two electrons occurs predominantly in the small area between the two nuclei. (The nuclei are at the locations of highest electron density). *Two electrons held together by two nuclei form a chemical bond between two hydrogen atoms in a hydrogen molecule.*

2.3. Physical constants of hydrogen

The diameter of a hydrogen molecule is 0.212 *nm* (2.12 Â). The dissociation energy of a hydrogen molecule into atoms by the reaction $H_2 \rightleftarrows 2H$ is 431.2 *kj/mol* (103kcal*!mol,* or 4.37 *ev/atom).* The thermal dissociation of gaseous hydrogen at temperatures below 2000°C is extremely small. Thus, in the temperature range 500-1500°C, the dissociation constant varies within [1]

$$k = \frac{(PH)^2}{(PH_2)} = 10^{-41}\ (225°C) \div 10^{-10}\ (1220°C)$$

The dissociation process can be activated by electric discharge, catalytic action of metals, various kinds of irradiation.

The diameter of a hydrogen atom, consisting of one proton and one electron, is about 0.1 *nm* (1 Â). The size of a proton is one hundred thousand times smaller than 10^{-6} *nm* (10^{-5} Â). The ionization energy of the free atom to the proton is 1307.2 *kJ/mol* (312 *kcal/mol,* or 13.53 *ev/atom).* When interacting with different reagents, hydrogen behaves differently, participating in covalent, ionic, and metallic bonds depending on the conditions [12].

As a consequence, hydrogen forms:

a) with alkali and alkaline-earth metals - salt-like hydrides, which are ionic crystalline non-volatile compounds;

б) with metals of group IVB, VB, VIB - hydrides, which are covalent compounds, gaseous at room temperature, e.g. AsH3 and SiH4;

в) in direct interaction with chromium, iron, cobalt, nickel, platinum, copper, silver, molybdenum, magnesium, aluminum, etc., the hydrides of some of these metals can be obtained only by preparative method; hydrides of some of these metals can be obtained only by preparative method. - true solutions; hydrides of some of these metals can be obtained only by preparative method;

г) with titanium, zirconium, thorium, hafnium, vanadium, niobium, tantalum, uranium, palladium, as well as with rare-earth metals in certain conditions and at concentrations -a - solid solutions; with increasing hydrogen content, there appear regions of mixtures of α - solid solutions with intermetallic compounds - hydrides, representing in most cases phases of variable chemical composition; finally, at even higher hydrogen content pure hydride phases are formed [18].

2.4. Thermodynamic regularities

The interaction of hydrogen with metals has been studied by many researchers, but the most complete description of these regularities is given in the monograph by Smitels, published for the first time in 1937 in New York and largely retained its significance to the present day

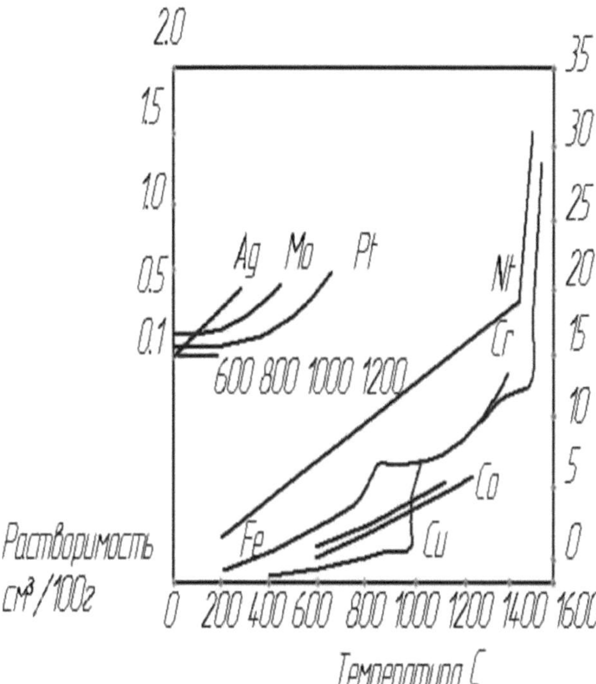

Figure 2. 2. Isobars of hydrogen solubility in iron, cobalt, copper, nickel, platinum, silver, aluminum, molybdenum [15, 85].

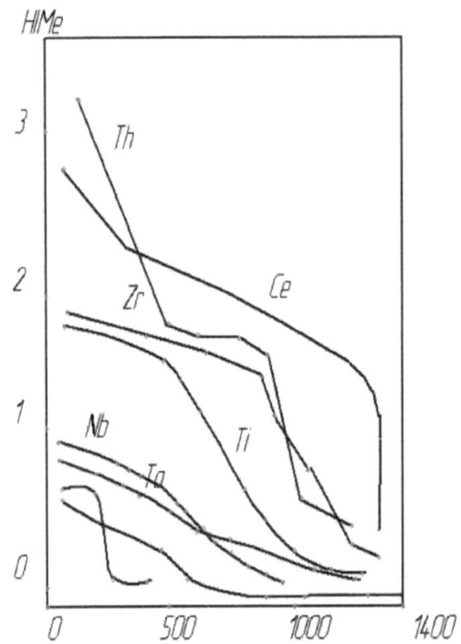

Fig. 2.3. Isobars of hydrogen absorption by palladium, vanadium, tantalum, niobium, titanium, zirconium, selenium and thorium.

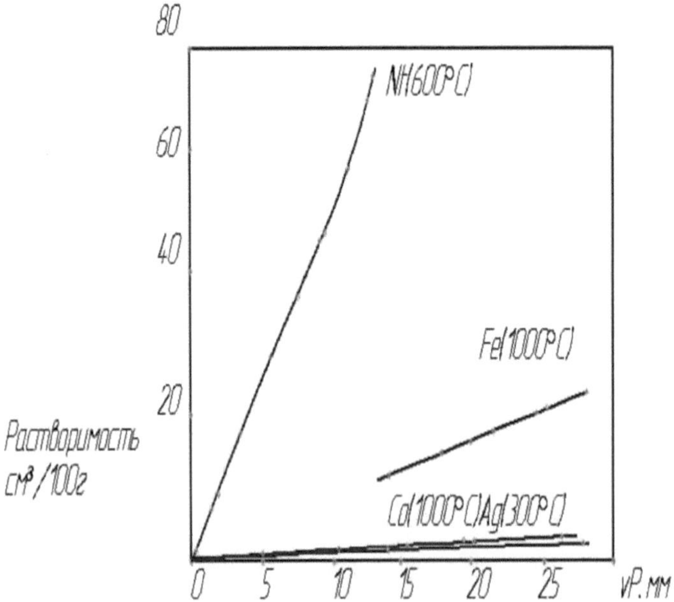

Fig. 2.4. Hydrogen solubility isotherms in nickel, iron, cobalt and silver [22]

Figures 2.2 and 2.3 show the absorption isobars, from which a sharp difference in the behavior of the metals can be seen. It follows from these figures that under equilibrium conditions with increasing temperature the amount of hydrogen absorbed by the metals of the grouping presented in Fig. 2.2 increases, and by the metals of the grouping presented in Fig. 2.3. decreases. The solubility of hydrogen in the metals of both groupings within each allotropic state increases with temperature. In the following, the term absorption, absorbed hydrogen is used to mean all the hydrogen absorbed by the metal, and the term solubility is used to mean the ability to absorb hydrogen within the formation of a true solid solution (H). The difference between the course of solubility (Fig. 2.3) and the change in the total amount of absorbed hydrogen (see Fig. 2.2) with increasing temperature in the metals of the grouping in Fig. 2.4. is due to the fact that at relatively low temperatures, metals such as titanium and zirconium can almost entirely convert to their respective hydrides. At high temperatures, the hydrides dissociate and only the hydrogen that forms the solid solution of introduction with them remains in the metals. As a result, the total hydrogen content of the metals decreases with increasing temperature. Figures 2.4 and 2.5 show the hydrogen absorption isotherms of some metals. The isotherms of nonhydride-forming metals obey Sieverts' law [19]:

$$\mathbf{S} = \mathbf{k_s}\sqrt{Ps}$$

where **S** is the concentration of hydrogen dissolved in the metal;

Ps-partial pressure of hydrogen in the gas phase above the metal

$\mathbf{k_s}$ is the solubility constant.

True dissolution, i.e. dissolution within the limits of solid solution formation, is an endothermic process. Following Sieverts' law at all temperatures and pressures achieved indicates that the metals of this grouping, i.e. chromium, platinum, copper, silver, molybdenum, magnesium and aluminum, form true solid solutions with hydrogen. It also follows from Siverts' law that hydrogen can dissolve, i.e. penetrate into the crystal lattice of a metal, only in the atomic state, i.e. dissociation of molecular hydrogen into atoms is obligatory before dissolution. The solubility of hydrogen in the metals of this grouping during the transition from the solid state to the liquid state increases up to a certain temperature, after which, with further increase in temperature, decreases to zero at the boiling point. The nature of the isotherms of hydrogen absorption by metals forming hydrides depends on temperature. At average temperatures, these isotherms have the form shown in Fig. 2.6. These are the isotherms of titanium, palladium, thorium, etc. (see Fig. 2.5). (see Fig. 2.5).

The rectilinear character of the isotherms of these elements at low pressures indicates the formation of solid solutions of introduction at these pressures. In the next pressure region, hydrides are formed, the amount of which increases with increasing pressure.

The third pressure region, where the absorption isotherm is parallel to the pressure axis (Fig. 2.6) corresponds to the saturation of metal hydrides. At prolonged exposure in this region, the metal phase completely disappears, giving way to hydride of limiting stoichiometric composition.

At higher temperatures, the appearance of the isotherm changes: the straightness is preserved in a wide concentration range, which is associated with the expansion of the region of existence of solid solutions at these temperatures [20, 21].

The formation of hydrides, as a rule, is accompanied by a large positive heat effect, i.e. the absorption of hydrogen by the metals of the grouping of Fig. 2.3. in most cases represents an exothermic process.

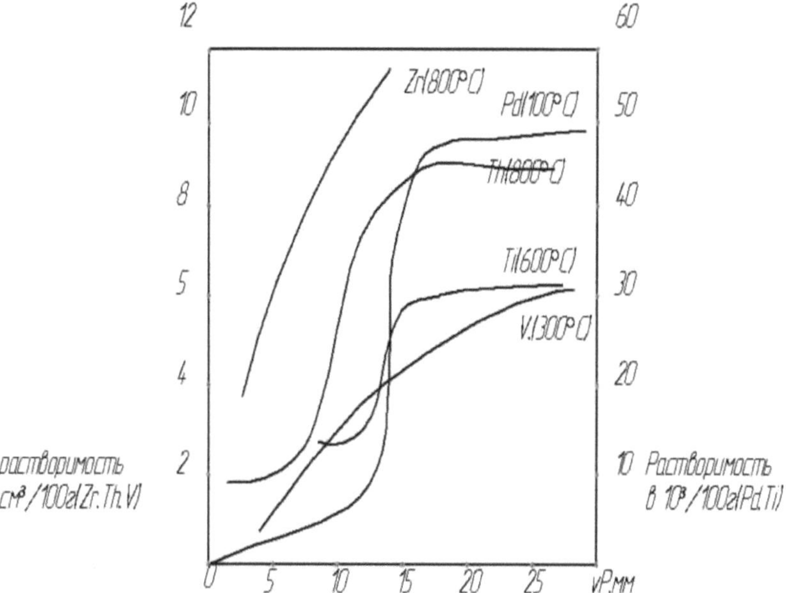

Fig.2.5: Hydrogen absorption isotherms of zirconium, palladium, thorium, titanium and vanadium[19].

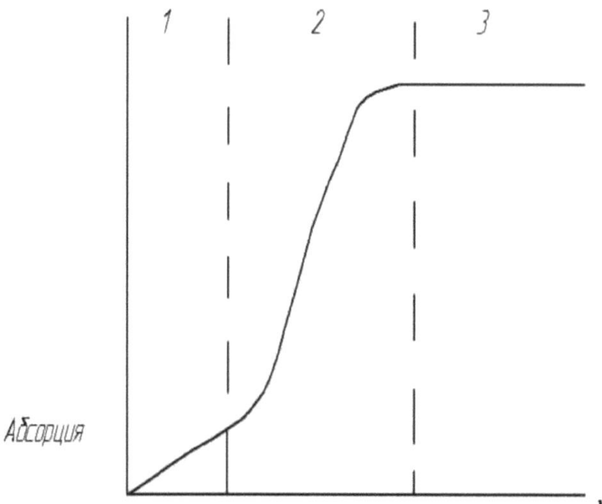

Fig. 2.6. Typical isotherm of hydrogen absorption (scheme) by hydride-forming metals at medium temperatures [23]

2.5. Kinetic parameters of interaction of metals with hydrogen

The interaction of metals with hydrogen is a complex process involving various reactions occurring on the surface and deep within the metal.

Reactions of metals with hydrogen can be divided into two broad classes - reactions that do not lead to the formation of chemical compounds on the surface, and reactions with the formation of these compounds. There are also reactions in which molecules of a chemical compound are formed on the surface, immediately passing into the gas phase.

Reactions of reversible absorption or release of hydrogen gas can proceed by a simple mechanism.

1. Mass transfer in the gas phase.
2. Adsorption of hydrogen molecules on the surface or its desorption G2- $2|G|_{ад}$.
3. Dissociation of adsorbed molecules and chemisorption of adsorbed atoms or recombination of adsorbed atoms into adsorbed molecules - G2 - $2G_{ад}$
4. Transition of adsorbed atoms to the adsorbed state Gad 2 $|G|_{т}$.
5. Hydrogen diffusion in metal.

Each of the successive steps of hydrogen uptake by the metal, with the exception of the second step, may be controllable depending on the conditions of mass transfer in the gas phase, surface condition, temperature, and the distribution of dissolved hydrogen concentration in the metal.

The rate of interaction between metal and hydrogen is conveniently represented as the density of the gas absorbed by the metal.

If the first stage of hydrogen supply to the surface is the controlling one, the process speed is determined by purely technical reasons and is proportional to the square of gas concentration in the metal.

$$J = k / c^2, \qquad (1)$$

where c is the concentration of gas in the metal.

If the controlling stage is the third stage then,

$$J = k_3\,(c^2_e - c^2), \qquad (2)$$

where c_e ***is the*** equilibrium concentration obtained from Sieverts' law

$$c_e = BP, \qquad (3)$$

Depending on the magnitude of the pressure P, the flow can be directed in either direction (absorption and emission).

This result is easily explainable, since in the case of the third controlling step one takes the number of dissociated molecules to be proportional to the number of molecules colliding with the surface, i.e., to the pressure, and the number of recombining atoms to be proportional to the number of their collisions, i.e., to the square of the surface concentration.

If the fourth step, i.e., the transition of adsorbed atoms to the dissolved state or the exit of dissolved atoms to the surface, is controlling, the flux density is proportional to the difference between the equilibrium concentration and the gas concentration in the metal.

$$J = k_4\,(c_e - c), \qquad (4)$$

Finally, in the case of diffusion control, when the concentration over the volume of the metal can no longer be considered constant,

$$J = -\,D\,c|_{x=0}, \qquad (5)$$

where D *is the* diffusion coefficient of dissolved gas in the metal.

Equations (1-5) are differential; in real processes there can be a transition from one controlling stage to another and periods of intermediate control when the rates of processes occurring at two different stages are comparable.

It should be recalled that the surface of the processed metal is very easily contaminated.

Consequently, the state of the workpiece surface will affect the 2nd, 3rd and 4th stages, i.e. the level of adsorption, and thus the level of reduction of surface energy of the deformed metal, as well as the rate of transition of adsorbed hydrogen atoms through the metal surface. When a layer of chemical compound appears on the metal surface, the reaction sites - metal and gas - appear separate from each other and their further interaction occurs only if at least one of the substances diffuses through the separating film. This leads to the fact that in many cases the reaction rate is determined not by the reaction itself, but by the processes of mass transfer through the film of the chemical compound. In the presence of cationic defects in the lattice of a chemical compound, metal ions diffuse to the film-gas phase interface and there interact with the gas. In the presence of anionic defects, the gas ions diffuse to the chemical compound-metal interface and interact with the metal at this interface.

Thus, both boundary reactions and transfer processes occur. The former include chemisorption of gas atoms and their transfer into the film lattice, as well as the transfer of metal and electrons into the film with the subsequent interaction of gas ions with metal at the metal-film boundary and metal ions with gas at the film-gas boundary. The latter include diffusion of metal cations and gas anions through the film along defect sites caused by the chemical potential gradient, and diffusion along grain boundaries of the chemical compound, penetration through pores and imperfections in it, as well as transport processes in thin films caused by spatial charges and electric field.

In real conditions, under sufficiently long experimental time, the transition from one dependence to another is possible, which is very smooth, which makes it difficult to analyze the curves and reveal the mechanism of gas absorption.

If the gas reacts not with pure metal, but with an alloy, the processes of internal formation of a chemical compound due to the preferential interaction of gas with one of the components of the alloy are possible. For multiphase alloys, the situation is even more complicated, since the formation of different phases of the compound and preferential interaction of gas with individual phases of metal is possible. There is also a redistribution of metal components of the alloy under the action of the chemical potential gradient occurring in the surface layer in the process of gas saturation [24 - 26].

2.6. Constants determining adsorption of hydrogen on metals

Adsorption is ideally understood as the enrichment of the interface between two immiscible phases by the components that make up these phases [27]. The interacting phases are always to some extent soluble in each other, so that adsorption should be understood only as an excess concentration of components at the interface compared to their content in the contacting phases.

The most important practical significance is the adsorption of gases on the surface of solids [28, 29]. Enrichment of the solid-gas interface by gases is caused by the unbalance on the surface between the atoms composing the solid, i.e., the unsaturation of bonds at the atoms lying on the surface. The saturation of these bonds as a result of gas adsorption leads to a decrease in the free surface energy Δ G and a decrease in the entropy ΔS, because when the position of the adsorbed particle on the surface is restricted, it loses some degrees of freedom. As a result, we obtain [27]: $\Delta G=\Delta H-T\Delta S$, where ΔH *is the* enthalpy change. Adsorption is always exothermic so that $\Delta H<0$. The heat of adsorption Q= - ΔH. The intensity of adsorption of gases is usually greater the stronger the chemical means between the metal and the interacting gas. Adsorption consists of two stages: physical and chemical. The former occurs at lower temperatures and is due to weak van der Waals forces. Chemical adsorption is preceded by dissociation of the molecule into atoms, which enter into a more or less close chemical bond with the surface atoms.

The difference between physical adsorption and chemical adsorption is that in the first process there is no transfer of electrons from gas molecules to the adsorbent, while chemisorption is associated with a significant redistribution of electrons between adsorbed particles (atoms) and the solid. However, when analyzing experimental data, it is not always possible to separate physical and chemical adsorption.

At the same time, the following criteria for distinguishing physical adsorption from chemical adsorption can be given:

1. The heat of chemisorption is much greater than that of physical adsorption and approaches the heat of chemical reactions. Thus, the heat of physical adsorption of hydrogen is usually less than 10 kJ/mol, and the heat of chemisorption of hydrogen is usually more than 60 kJ/mol.

2. The physical adsorption of a gas is very similar to its condensation, and so it occurs only at temperatures near or below the boiling point of the adsorbate at a given pressure.

The transition of gas molecules from the physically adsorbed state to the chemisorbed state occurs, as a rule, only in active centers, which are steps, corners, protrusions and depressions on the crystal surface, as well as lattice defects (vacancies, dislocations, grain and subgrain boundaries), In the state of physical adsorption molecules move freely along the surface, while in chemisorption adsorbed atoms (or partially dissociated molecules) can be mobile or immobile depending on the height of potential barriers and temperature.

The efficiency of chemisorption is characterized by the adhesion coefficient s, which is defined as the number of chemisorbed particles relative to those colliding with the surface. Thus, in contrast to the condensation coefficient α which reflects the efficiency of physical adsorption, the sticking coefficient s refers to chemisorption. The condensation coefficient α is close to unity even for systems with low heat of adsorption. This is due to the fact that only collision is sufficient for condensation of gas molecules by the physical adsorption mechanism. In contrast, the sticking coefficient is much smaller than unity (~0.1), since chemisorption requires suitable sites (free chemisorption centers).

At chemisorption of hydrogen on metals three types are distinguished: *A* , *B* and *C* (Fig. 2.7) [30]. Adsorption type *C is* observed at low temperatures (below -100 °C). It is characterized by a relatively small heat of adsorption (21 - 63 kJ/mol). However, it is not a physical adsorption because the heat of adsorption of type *C adsorption is* greater than what is considered reasonable for physical adsorption. Type *A* adsorption is a typical strong chemisorption with a fairly large heat (Figure 2.7); it is observed at temperatures close to room temperature. At higher temperatures, type *A* adsorption is replaced by type *B* adsorption, another type of strong chemisorption. The two types of strong hydrogen chemisorption can be attributed to the different nature of the adsorbed particles, viz: H^+ 2 and H^- [31].

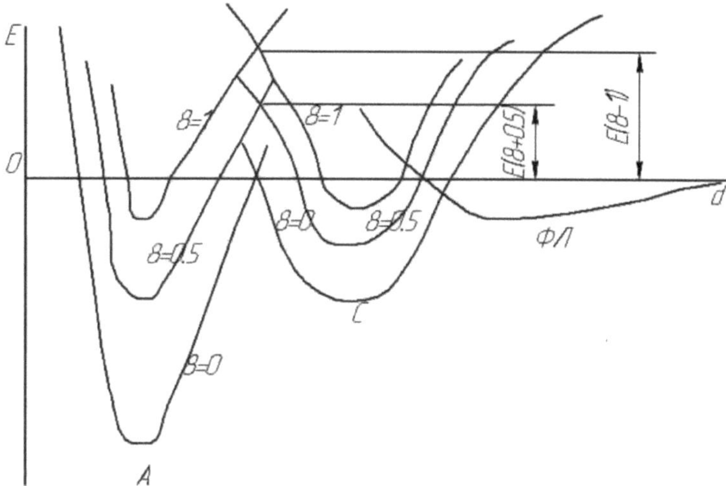

Fig. 2.7. Scheme of energy change in hydrogen adsorption on metal with increasing degree of coverage 0 at adsorption of different types (A,C,FA - physical adsorption) [30].

2.7. Diffusional mobility of hydrogen in metals

The diffusion mechanisms of embedded atoms depend significantly on temperature, and the diffusion mechanism of hydrogen differs significantly from those of other embedding impurities [24, 25]. This is explained by the fact that after ionization, oxygen and nitrogen atoms have sizes comparable to those of the internodes, while hydrogen is in the lattice as a proton quasion screened by the gas. In addition, at low temperatures, a significant contribution to hydrogen diffusion is made by proton tunneling transitions from one internode to another [32].

Hydrogen in metals has an unusually large diffusion mobility. Even at temperatures below room temperature, it is large enough to redistribute in volumes comparable to the size of micrograins. At room temperature, the diffusion coefficient of hydrogen in ferrite is 1010 - 10^{11} times greater than that of carbon and nitrogen, and 10^{35} times greater than typical values of diffusion coefficients of substitutional elements.

At temperatures close to absolute zero, hydrogen atoms are not localized in the lattice and form zone states if they are not trapped by lattice defects. At these temperatures, the movement of hydrogen atoms in the lattice is accomplished by coherent tunneling. This is only a theoretically expected mechanism: so far no coherent tunneling has been found in the diffusion of hydrogen in metals [32].

At higher temperatures hydrogen atoms are localized in a certain internode or near it and diffusion is performed by thermally activated jumps of atoms from one internode to another. An atom can jump from one equilibrium position to another either by tunneling or as a result of acquisition of additional energy sufficient to overcome the potential barrier.

At sufficiently high temperatures, especially in o.c.c. metals, hydrogen atoms so frequently jump from one equilibrium position to another that the average time between two consecutive jumps becomes comparable to the duration of the hydrogen atom's stay in the internode [32]. Under such conditions, it makes no sense to make a distinction between resting states to overshoots. In this case, the diffusion of hydrogen atoms occurs approximately as in a dense gas or liquid. This mechanism of diffusion of introduction atoms is called liquid diffusion [32]. Naturally, there are no clear temperature boundaries in which one or another diffusion mechanism operates.

All of the above refers to the diffusion of the introduction impurity in an ideal lattice. The diffusion coefficient, which determines the rate of this process, is called the true or lattice diffusion coefficient. In a real metal, however, there are crystalline imperfections: grain and phase boundaries, particles of the second phase, pores, which have a significant influence on diffusion processes. This influence can have a twofold character. First, dislocations, grain and phase boundaries, elongated particles of the second phase can serve as paths of preferential diffusion due to a significantly higher diffusion coefficient compared to the lattice of the base metal [33]. Secondly, imperfections can be collectors or traps for the introduction impurity due to the energetic advantageous location of the impurity in the imperfections themselves or in the distorted lattice near them. Depending on the binding energy of the impurity to the trap, traps are divided into irreversible traps, from which the impurity is practically not released, and reversible traps, from which the impurity is released at a sufficiently low concentration in the surrounding solid solution. The efficiency of traps increases significantly with decreasing temperature.

As a rule, diffusion processes in real metals are attempted to be described by diffusion equations, in which the effective diffusion coefficient is used - a coefficient that approximates the averaged influence of traps and paths of preferential diffusion on the processes of diffusion movement of impurity. If such approximation turns out to be too coarse, then it is necessary to solve the problem of diffusion in heterogeneous material, modeling traps by sources and sinks of impurity, which considerably complicates calculations.

There are two main approaches to diffusion processes: kinetic, based on the calculation of displacements of chaotically moving impurity atoms, and phenomenological, based on the provisions of thermodynamics of irreversible processes. In most cases, both of these approaches give similar results, but the phenomenological approach is used much more often due to the greater simplicity of calculations.

2.8. Processes and phenomena responsible for facilitating deformation and fracture of metal in hydrogen

The study of phenomena occurring during the interaction of hydrogen with deformed metal is of great interest not only for the protection of metals from undesirable consequences of these processes, but also for the scientifically justified use of the effect of this interaction in order to obtain the necessary positive result. Knowledge of the physical essence of the processes occurring in the interaction of deformed metal with hydrogen and the creation of a theory of this phenomenon is an urgent task.

Of the wide range of questions included in the discussed problem, the most important are the questions of the mechanism of change of deformative properties of metals, primarily iron and its alloys, at reversible interaction with hydrogen.

Recently, the influence of the essence of this mechanism has become even more important in connection with the emergence of a new generation of polymer-based POTS with extraordinary efficiency, which is associated with the influence of hydrogen, which is formed in the zone of mechanical processing under the influence of high temperature due to pyrolytic reactions of polymer components of POTS.

Recently obtained experimental data on the behavior of metals deformed in gaseous hydrogen [34] at room temperature and atmospheric pressure allowed us to consider this problem from the standpoint of the physicochemical theory of deformation and fracture of solids created by Academician P.A. Rebinder [35,36] and the "unified hypothesis" of hydrogen embrittlement proposed by G.V. Karpenko [37,38].

The hypothesis of the change in deformational properties of metals proposed here, in contrast to the known adsorption hypothesis [39], is based on the local reduction of interatomic bond forces in the most overstressed microvolumes with broken and uncompensated bonds as a result of the act of hydrogen chemisorption. Due to the reduction of interatomic bond forces, their rearrangement and rupture is facilitated. The change of certain properties of metal as a result of its formation in interaction with hydrogen is predetermined by the degree of heterogeneity of deformational properties of its microvolumes, type of stress state, strain rate, specificity of temperature dependence of hydrogen adsorption on metal and selective character of chemisorption.

In the physicochemical theory of deformation and fracture of solids, there is a semi-empirical rule Rebinder-Shukin-Lichtman, according to which a surfactant must have: a) limited but finite solubility; b) sufficient mobility; c) a certain affinity for a given solid. Complete absence of solubility indicates insignificant chemical affinity, and at high solubility the surface migration of atoms, responsible for the reduction of surface energy, will be inhibited by intensive bulk diffusion deep into the solid.

The iron-hydrogen system fully complies with all points of this rule. The true solubility of hydrogen in the iron crystal lattice (equilibrium concentration) at room temperature and normal pressure is only 5 - 10^{-8} weight % [40]. This solubility

differs from the total, which includes hydrogen released in the form of excess forms of various defects and non-plosions of the metal structure.

The size of the hydrogen atom is minimal, so hydrogen easily penetrates into submicrodefects of the structure. The total diffusion coefficient, reflecting under normal conditions the rate of penetration mainly along grain boundaries, is quite high (10^{-7} cm^2 /sec), which indicates the great mobility of hydrogen. Due to the specific electronic structure of the atom, hydrogen has metallic properties, which is the reason for its affinity with metals. Thus, low solubility, high mobility and affinity with metals allow to characterize hydrogen as the most effective surfactant in relation to metals.

The binding energy of hydrogen and iron atoms is equal to the sum of the dissociation energy of molecules (104 *kcal/mol*) and the heat of chemisorption (32 *kcal/mol*), i.e., 136 kcal/mol. Having such a high binding energy, chemisorbed hydrogen reduces the bonding forces of metal atoms, as evidenced by the experimentally observed increase in the lattice parameter [41, 42], as well as the breakage of metallic and the formation of covalent bonds [43].

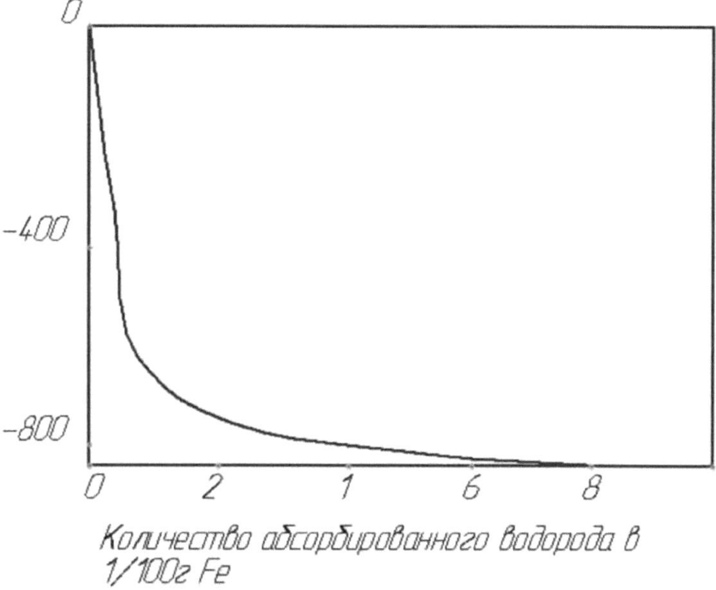

Figure 2.8. Decrease in surface energy of iron caused by hydrogen adsorption at 20°C [11]

However, in the case of adsorption, the summands of this sum cannot be considered as independent variables. Numerous studies by P. A. Rebinder and his school [35, 36, 45, 46] have shown that a decrease in surface energy due to adsorption facilitates plastic deformation, i.e., reduces the work of plastic deformation. Consequently, in the above relation, the work of plastic deformation p = p (γ_0) and at $\gamma_0 \to 0$ the value of *p* also tends to zero [47].

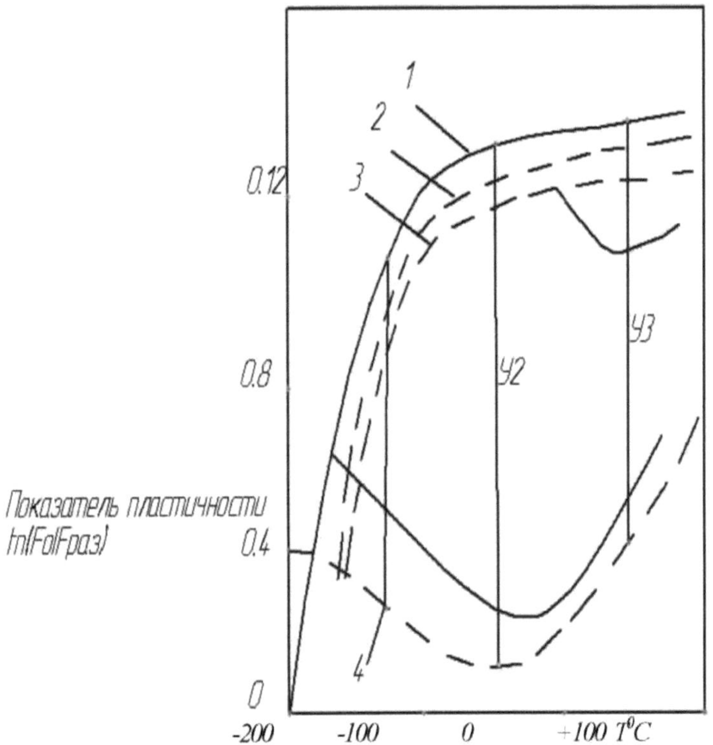

Fig. 2.9. Dependence of plasticity indices of low-carbon steel on temperature and strain rate [30]; 1, 2, 3 - samples not saturated with hydrogen; II, III - samples cathodically saturated with hydrogen; strain rate 5% /min (1, 1), 10^4 %/min. (2, II), 5 - 10^5 %/min (3, III); 4 - theoretical curve of change of plasticity index of iron and its alloys under the influence of hydrogen under the condition of commensurability of deformation rate with the rate of chemisorption and migration.

In this approach to the physical nature of the phenomenon, φ_0 turns out to be a multiple of Y_0 , i.e., a decrease in γ_0 in the presence of an adsorption-active element leads to a proportional decrease in F_0 Thus, a decrease in surface energy can significantly affect the overall work of destruction.

It is known that the effect of hydrogen on crack growth in quenched and tempered steel is essentially the same when exposed to gaseous hydrogen, hydrogen produced during stress corrosion and released during electrolysis on the metal before or at the time of testing [48]. Therefore, when considering the mechanism of hydrogen influence on mechanical properties, we will focus on general patterns rather than on special cases, without making distinctions between hydrogen sources.

The complex nature of the dependence of mechanical properties of metals interacting with hydrogen on deformation rate temperature is well known. At constant strain rate, hydrogen intensively reduces plasticity indices (as well as other deformative properties) only in a certain temperature range, i.e., a minimum is observed on the temperature dependence curve (Fig. 3.9) [49]. As the strain rate decreases, the degree of change in mechanical properties increases and the minimum shifts to lower temperatures. The explanation of these experimentally observed temperature-velocity dependences is a criterion for assessing the validity of a hypothesis concerning the phenomena occurring in the interaction of metal with hydrogen.

It is known [50] that physical and chemical adsorption decreases with increasing temperature, but the transition from the first to the second is accompanied by an increase in the amount of gas adsorbed by the metal surface; as a result, a maximum appears on the adsorption isobar in a certain temperature range characteristic of the metal (Fig. 2.9). It was previously shown [51] that the greatest changes in the mechanical properties of the metal occur precisely in this temperature range corresponding to the maximum on the adsorption isobar.

In accordance with the considered hypothesis, the efficiency of hydrogen action on mechanical properties is determined by the rate of migration of adsorbed atoms into microvolumes of local shear regions of deformed metal and the amount of gas chemisorbed at a given temperature. At any test temperature, the effectiveness of hydrogen decreases with increasing strain rate. This is due to the fact that at a given temperature the rates of hydrogen chemisorption and migration are constant, so that as the deformation rate increases, both the total amount and depth of hydrogen penetration along the planes of facilitated sliding decreases.

The adsorption isobar shows that the amount of chemisorbed gas increases with increasing temperature (in the interval of transition of physical adsorption to chemical adsorption). However, when the maximum chemisorption is reached, the total amount of chemisorbed gas decreases with increasing temperature. If the strain rate were commensurate with the rate of chemisorption and surface migration, the efficiency of hydrogen action at any temperature would be determined only by the amount of chemisorbed gas. At preservation of such commensurability the curve of temperature dependence of strength properties of the metal interacting with hydrogen would mirror the adsorption isobar (Fig.3.9 curve 4). The maximum efficiency of hydrogen influence found at some constant strain rate (Fig. 3.9) is due to the optimal combination at this temperature of the amount of chemisorbed gas per unit time (i.e., chemisorption rate, migration rate, and given strain rate, see Y_1, Y_2, and Y3 in Fig. 2.9 and Fig. 2.10).

The action of external forces, always aimed at breaking the existing interatomic bonds in the metal, is opposed to the ability of the metal to resist shape change (strength) and the ability to restore broken bonds, while changing the shape (ductility). Reducing the forces of interatomic bonds due to hydrogen chemisorption will facilitate both their rupture and rearrangement, i.e., reduce strength and increase ductility.

Depending on the conditions, type of deformation and mechanical properties of the metal, irreversible breaking of interatomic bonds will prevail or

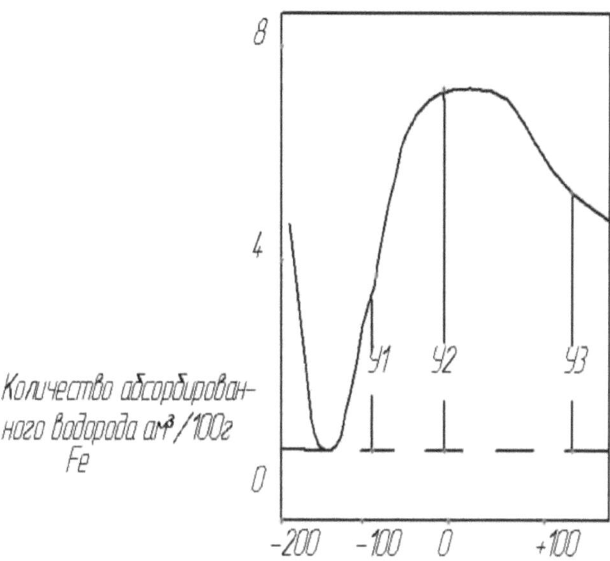

Fig. 2.10 Isobar of hydrogen adsorption on iron (scheme) [50]

rearrangement of bonds, so the influence of hydrogen can manifest itself in various forms.

In metal tensile fracture is always preceded by plastic deformation, which prepares the crack nucleus. In polycrystalline metal, which is a heterogeneous system, plastic deformation is localized in the most overstressed microvolumes. Due to the reduction of surface energy by chemisorbed hydrogen, plastic flow in these microvolumes begins at lower stresses, which is the reason for the decrease in the upper [52,53] and lower [54, 55] yield strengths observed in saturated tensile stresses.

by iron hydrogen. At the same time, penetrating into the area of shear localization along submicrodefects in slip planes, hydrogen facilitates the rupture of interatomic bonds under the action of normal stresses, so cracks appear [56] and develop at lower stresses [57, 58]. Ultimately, such localization of fracture due to, as will be shown below, selective chemisorption leads to quasi-brittle along slip planes or intergranular fracture [59] and to a decrease in the relative contraction [37] of the deformed specimen.

If the metal is deformed under conditions that impede plastic flow (after heat treatment or under stress), then in the presence of hydrogen, the work expended on elastic stretching may be sufficient to break interatomic bonds at the crack tips [58]. If, however, deformation is carried out under conditions that restrain the formation of new metal separation (fracture) planes, for example, under the action of compressive stresses, then the reduction of interatomic forces will facilitate the rearrangement of atoms (i.e., shape change) under the action of external forces. In this case, hydrogen adsorption will facilitate the flow of plastic deformation due to the reduction of the surface potential barrier, facilitating the exit of dislocations to the surface, nucleation and generation of subsurface sources. This facilitates the movement of dislocations and increases their density, resulting in more intense strain hardening.

Thus, the final result of deformation of iron and its alloys in the presence of hydrogen depends on specific deformation conditions and can be different and even opposite.

The diversity of effects observed during deformation of a metal interacting with hydrogen is due not only to deformation conditions, but also to some peculiarities of gas adsorption on the metal. Due to the heterogeneity of the metal, hydrogen adsorption on its surface occurs selectively, mainly in places with maximum free energy, i.e. on defects in the structure, at the dislocation exit sites, intergranular boundaries, in slip planes, near inclusions, etc. Adsorption activity of metal increases with increasing number of defects formed during deformation and playing the role of active centers. The presence of defects in the metal leads to the fact that deformation and fracture, facilitated by hydrogen adsorption, are localized in certain micro- and submicro-volumes. In turn, this causes a very intensive localization of adsorption in submicro-volumes, which eventually, under tensile stresses, leads to quasi-brittle (reduction of relative contraction and elongation) fracture at much lower energy costs. If the number of defects is small and the structure of the metal is close to that of an ideal crystal, hydrogen adsorption is not accompanied by significant changes in the mechanical properties of the metal. Hydrogen does not cause quasi-brittle fracture of iron single crystals [60-62] and zone-melting iron [63]. This is due to the insignificant stationary activity of the metal and the homogeneity of the deformation properties of its microvolumes, which makes it difficult to localize the deformation. On the other hand, a metal containing a very large number of defects, although it has an increased adsorption activity, is more homogeneous in its adsorption and deformation properties [31]. In this case, when deforming the metal in the presence of hydrogen, the strain is distributed more uniformly throughout the volume and can contribute to increasing the plasticity reserve of the metal.

When hydrogen interacts with the deformed metal under certain specific conditions, a number of related processes occur: formation of collectors, interaction of hydrogen with impurities and inclusions, formation of new phases (hydrides), etc., as a result of which internal stresses arise in the metal, formation of non-continuities, etc.. These processes also have a significant impact on the deformative properties of the metal (for example, decarburization at high temperatures will predetermine the change in mechanical properties). However, primary in all cases is the act of adsorption, which dramatically changes the deformational properties of the metal even in the absence of concomitant processes. Therefore, the main attention in this work is focused precisely on those changes that occur as a result of interatomic interaction of chemisorbed hydrogen with the metal matrix.

The analysis of the considered phenomenon from the positions of physicochemical theory of deformation and fracture allows to determine the ways of metal protection in those cases when the interaction of hydrogen with metal leads to negative consequences, and ways of using this interaction to obtain a positive effect in a number of industrial technologies and in operational conditions.

Hydrogen is adsorbed on all metals and, depending on the magnitude of the resulting interatomic bonding forces, affects mechanical properties to a greater or lesser extent. Therefore, protecting metals by coating them with other metals that are more resistant to hydrogen is a half-measure.

The most promising for the protection of metal subjected to cyclic or static tensile stresses are elastic polymer coatings preventing hydrogen contact with juvenile metal surfaces [64, 65], and at higher temperatures - special oxide films [66] created on metal by a certain technology. A significant reserve for increasing serviceability in hydrogen-containing environments is the reduction of contamination of metals with non-metallic inclusions (especially sulfides, string aluminosilicates, alumina oxides, etc.). [64], as well as the use of various treatments that homogenize the deformative properties of metal microvolumes (riveting, pre-deformation, thermomechanical treatment, etc.) [64-71].

The proposed mechanism also allows us to phenomenologically justify the possibility of using hydrogen as a technological medium facilitating deformation and fracture.

The results of studies have shown that during compression deformation (roller rolling, shot blasting and other types of riveting) the reduction of interatomic bonding forces in the near-surface layer due to hydrogen chemisorption facilitates plastic flow and leads to more intensive hardening. It has been experimentally established that during metal pressure treatment (cold rolling, calibration, etc.) with the use of hydrogen as a process medium, the deformation force is significantly reduced and the surface cleanliness class of the metal increases [72]. Similar results have been obtained in metal cutting processing [73]. In all types of machining, the presence of hydrogen (released in one way or another) in the metal-tool contact zone is necessary to achieve a positive effect. [73,74].

Since the discovery by P. A. Rebinder of the effect of adsorption facilitation of deformation and fracture of solids, a huge amount of experimental material has been accumulated on the changes in the deformation properties of metals in various organic surfactant media. It has been found [73] that these changes are proportional to the molecular weight, i.e., the number of carbon atoms in the molecule of the surfactant organic compound. In the previous chapters of the work, experimental facts were presented, indicating that the efficiency of the surfactants based on high molecular weight compounds is explained by the fact that the thermal destruction of the polymer base of the surfactants leads at the final stage to the formation of hydrogen in the active form and its influence on the process of deformation and destruction of metal, i.e. on the formation of the sheared layer.

In terms of the considered hypothesis of hydrogen facilitation of formation and fracture, this fact is of considerable scientific and practical interest, since it is possible to scientifically substantiate the issue of creation and selection of lubricants and lubricating and cooling fluids, guided by their ability to dehydrogenation in specific conditions of operation or processing of machine parts. In addition, the observed regularity of increasing influence of organic compounds on deformation properties of metal with the increase of their molecular weight is explained.

When creating and selecting lubricants for friction units, the ability of organic compounds to dehydrogenation is of special importance. It has been experimentally established that hydrogen release on the friction surface contributes to faster working-in of parts, eliminates scoring under overloading and reduces the friction coefficient almost 2 times. The considered hypothesis also explains the phenomena of anomalously low friction coefficient after radiation irradiation of metal-polymer friction pair [74], as well as the phenomenon of selective mass transfer, which is observed at introduction of surface-active additives [75] serving as a source of hydrogen.

The analysis of the whole variety of effects caused by the interaction of hydrogen and hydrogen-containing media with deformed metal, carried out from the standpoint of physical and chemical theory of deformation and fracture, allowed to find an explanation not only for the established facts, but also to put forward new proposals in scientific and practical aspects.

The above presented research results confirm the fruitfulness of the considered hypothesis.

So, the essence of the considered phenomenon can be briefly formulated as follows: it has been experimentally discovered that in deformation processes of metals in contact with hydrogen, which is released from hydrogen-containing media due to dehydrogenation of their molecules, there is a phenomenon of hydrogen facilitation of formation and destruction, which consists in the fact that hydrogen, chemisorbed on the external and internal surfaces activated by deformation, weakens interatomic bonds of the metal, facilitating their rupture and rearrangement; depending on the conditions of deformation of metals in hydrogen-containing media. This phenomenon is observed in numerous operating conditions and technological processes during mechanical processing of metals in hydrogen-containing media.

Chapter III

Processes of adsorption of chemical elements, included in the composition of SOTS, into the treated metal.

3.1 Adsorption of COTC pyrolysis products

3.1.1 Adsorption at the interface

Cutting of steels and alloys is considered to be a process of prevailing plastic deformation [76]. The thermal regime and specific loads on the working surfaces of the tool, and consequently, the intensity and nature of its wear depend on plastic deformation. Machining accuracy, roughness and quality of the surface layer of the machined part depend on the nature of plastic deformation and the mechanism of strain hardening. Therefore, in order to significantly reduce the energy and force costs of metal machining, it is necessary to use SOTS, which, in addition to lubricating and cooling functions, would have an impact on the value of energy costs, i.e. on the work of plastic deformation. However, the main principles underlying the currently developed domestic and foreign coolants [77-79] are mainly aimed at solving the problem of increasing their antifriction properties, ensuring with their help the temperature reduction in the zone of mechanical processing. Such indicators of properties of technological fluids, although important, do not solve the main task - reduction of energy and power costs of machining process. It should be noted that a certain influence on facilitating the process of fracture of modern coolants is due to reversible effects as a result of adsorption of low molecular weight components. However, the complex effect of these components, included in the composition of modern coolants, on the process of mechanical treatment of solids is still not significant and does not provide a high level of requirements to such liquids.

The data presented in the previous paragraph indicate that low-molecular-weight products formed as a result of depolymerization of the macromolecular chain of the polymer in the zone of mechanical processing are mainly of hydrocarbon composition.

Consequently, it is such media that most significantly influence the process of plastic deformation, significantly reducing energy costs during mechanical processing of metals in polymer-containing SOTS (Fig. 3.1).

The influence of low-molecular hydrocarbon media on the process of plastic deformation of metal is realized through its surface, by physical or chemical adsorption. In this regard, the results of the study of adsorption surface of active pyrolysis products of polymeric component of SOTS on the treated metal surface by cutting or OMD are presented below.

On the surface of steel 45 after turning in air, a large concentration of oxygen atoms is observed (Fig.3.1.), which is evidence of intensive oxidative processes on iron initiated by high temperature in the processing zone. In addition, a high carbon content was found on the surface of the sample. High concentration of oxygen in the surface layer of steel processed in the air is due to the fact that in the space between the cutting surface and the main back surface of the tool penetrates the air medium. The air oxygen continuously oxidizes the thin surface layer of the tool material and the cutting surface. As for the carbon found on the machined surface, the high atomic concentration is attributed to the fact that during the cutting process, there is a continuous and directed diffusive transfer of carbon from the boundary layer of the tool blade to the contacting layer of the machined material [80]. At a depth of about 20 nm, the atomic concentration of chemical elements in steel is different.

carbon and oxygen signals practically disappear, and the atomic concentration of these elements is at the level characteristic of deep layers of steel 45.

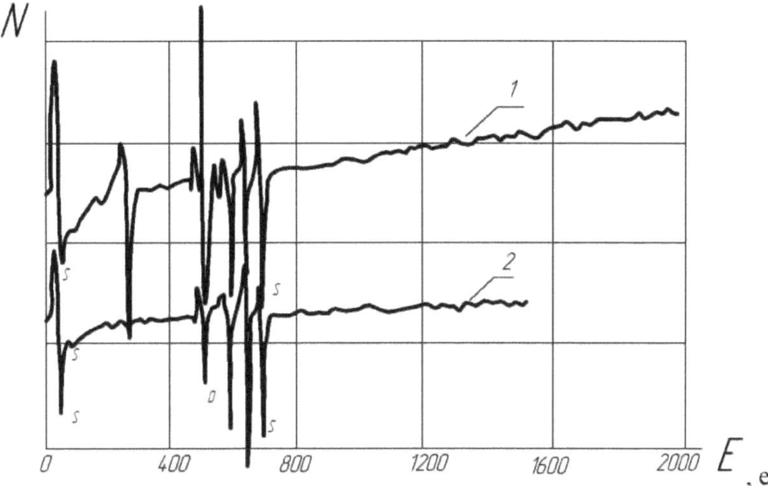

Fig.3.1. OLE spectra taken from the surface of steel 45 after end turning in air: 1 - surface; 2 - at a depth of 20 nm. Cutter VK - 6; cutting mode: n = 8.3 s^{-1} , t = 0.8, S = 0.1 mm/rev. HRC 42...45. The samples after processing were subjected to ultrasonic cleaning for 7 min and drying in an inert atmosphere at 125°C for 30 min.

When turning the same steel in tap water, traces of S and C1 are detected on the surface of the sample, unlike those processed in air, and the concentration of oxygen decreases (Fig.3.2.). This appears to be due to the adsorption of C1 and S from water that contains these chemical elements and the chemical interaction of oxygen dissolved in water with iron. Since the concentration of atomic oxygen in water is much lower than in air, its content on the surface of the treated metal is also lower. The distribution of chemical elements in the volume of the material completely coincide with the first considered case, when the material was treated in air.

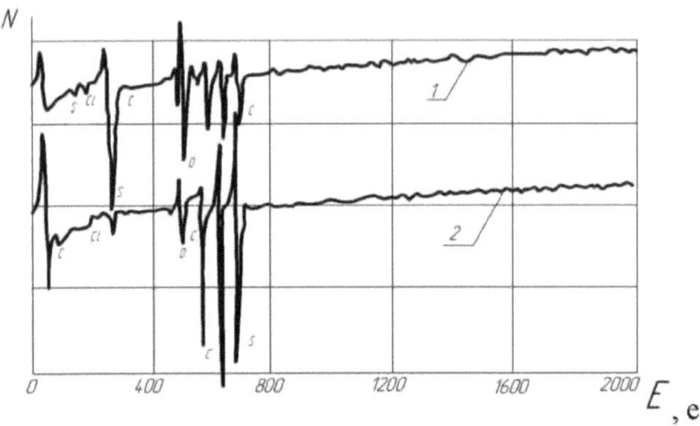

Fig. 3.2. OLE spectra taken from the surface of steel 45 samples after turning in water: 1 - surface; 2 - at a depth of 20 nm.

When PVC is added to water (3% by dry residue), the OGE spectrometer registers the signal of element S, as for water, strong signals characteristic of C1 and C, weak signal of O and practically no signal of Fe (Fig.3.3.). The high concentration of C1 and C on the sample surface indicates depolymerization of the polymer macrochain by the mechanism described in the previous paragraph. Practically complete absence of oxide films on the surface confirms the fact that hydrogen and hydrocarbon compounds formed in the process of depolymerization of the macrochain completely binds oxygen, i.e. works as a reducing medium.

Thus, a layer, about 20 nm thick, consisting of C1 and C elements, the depolymerization products of the polymer macrochains, is formed on the steel surface.

At etching of the adsorbed layer the C1 signal disappears, the C concentration decreases, the O concentration remains the same as on the surface and the Fe signal appears. It should be noted that when steel is treated in commercial emulsol ET-2, the same chemical elements are adsorbed on the treated surface and in the same concentration as in tap water (Fig. 3.4***)***. Hence, two important conclusions follow - the observed efficiency of emulsol ET-2 is not related to the decomposition of elements contained in it, and adsorption of these elements is reversible, i.e. in this case there is a physical adsorption of components from the emulsol.

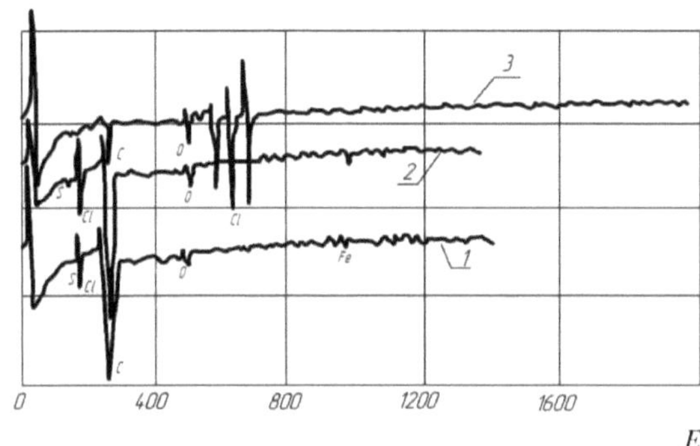

Figure 3.3. OLE - spectra taken from the surface of steel 45 after turning in water with PVC:

1, 2 - surface, 3 - at a depth of 20 nm

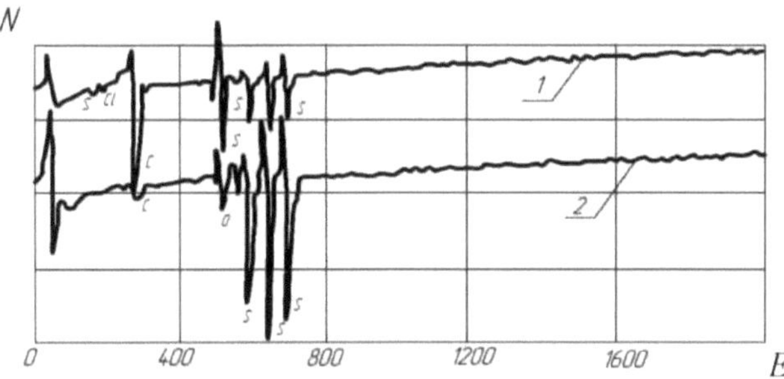

Fig.3.4. OLE - spectra taken from the surface of steel 45 after flowing in emulsol ET-2: 1 surface; 2 - at a depth of 20 nm.

On the surface of samples from technically pure metals, there is a large concentration of carbon (up to 60-80 at. %), weakly depending on the nature of the liquid used during processing and on the type of processed metal (Fig.3.5)

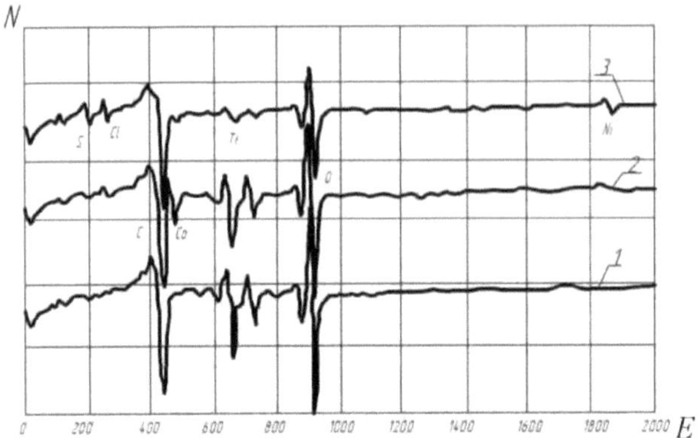

Fig. 3.5. OLE - spectra taken from the surface of technically pure titanium after flow: 1 - in air; 2 - in water; 3 - in the same water with PVC (3% by dry residue).

Figure 3.5. shows only the OLE spectra of Ti, since the spectra of other studied metals Ni, A1, Co, Si, Cg are qualitatively similar and differ only in the oxygen content.

Significant enrichment of the surface of titanium samples with carbon is explained, as well as for steel, by the directed diffusion transfer of carbon from the surface layer of the tool into the processed metal in contact with it.

On the surface of pure metal samples different oxygen content is observed from 5 at.% for Ni and Cr, to 30-50 at.% for Ti, which can be explained by different chemical activity of these metals towards oxygen.

The conducted studies show that the efficiency of commercially produced by the industry of commercially available coolants and investigated technological means is mainly conditioned by physical processes developing between the contact surface of the cutter and the workpiece, and in metal pressure treatment - between the contact surface of the forming tool and the deformed surface. However, during machining in media containing high-molecular substances, a more complex complex complex of thermomechanochemical transformations of the polymer macrochain with the formation of various chemically active products, mainly of hydrocarbon composition, takes place [80]. These products, chemisorbing on the surface of the treated metal, contribute to the reduction of its surface energy, facilitating, according to the Rebinder effect, the processes of deformation and destruction, as well as the formation on the surface of the material "pile" consisting of fragments of the polymer macrochain, causing a decrease in the coefficient of friction. In addition, due to a stronger "metal-macroradical" bond, the probability of macroradicals getting directly into the contact zone "workpiece - tool blade" increases.

Depolymerization of macrochains to form gaseous, liquid and solid low molecular weight products causes the following processes:

1. Reduction of oxygen concentration in the cutting zone during turning or in the deformation zone during metal forming during metal forming, creates conditions for cutting or forming tools to work in a reducing atmosphere, which is known to increase their wear resistance.

2. Adsorption of HC1 formed by the interaction of hydrogen and chlorine and exposure of the acid to the steel surface to form the unconfined salt $FeCl_2$, and when oxidized, $FeCl_3$. The melting point of these salts is low and does not exceed 672°C and 309°C for $FeC1_2$ and $FeC1_3$, respectively. Therefore, during machining in which the temperature in the cutting zone is much higher than the melting point of inorganic salts, they melt, forming films with low shear resistance, which reduces the coefficient of friction.

3. Adsorption of atomic hydrogen formed as a result of depolymerization of the polymer macrochain and after interaction with the treated surface by the reaction Fe + 2NS1 = $FeC1_2 + H_2$, followed by dissociation. The effect on the metal of hydrogen adsorbed on its surface is extremely diverse and the mechanism of its influence is not completely clear. However, in all cases considered, it facilitates deformation and fracture processes.

4. Adsorption of other chemical elements that are or can be specifically introduced (e.g., by modification) into the composition of macromolecules and, primarily, C, H, P, S, and others. These adsorbed substances can fulfill various functions. For example, to directly separate the rubbing surfaces (carbon), to separate the rubbing surfaces after the formation of chemical compounds with iron.

5. Diffusion processes of all chemical elements adsorbed on the surface of the treated material, with changes in its physical and mechanical properties.

Thus, mechanical processing of steel and alloys (by cutting or by pressure (OMD) in polymer SOTS is carried out under the complex and diverse influence of various chemical elements and compounds, the combination of which is determined by the chemical composition of the polymer macrochain, depolymerization conditions, catalytic activity of the surface of the treated material and its properties. The variety of possible manifestations of phenomena and processes leads to a significant difference between the mechanism of metal cutting and pressure treatment in media based on polymers and SOTS based on low molecular weight components.

3.1.2 Adsorption localized in the molecular layer of deformed metal

It is known that, in general, the effect of high temperature and mechanical loads on the polymer can cause a change in the chemical composition of the polymer links of the polymer component of the SOTS (or model medium), a change in the multiplicity of bonds, rearrangement of atoms in the macromolecule, the appearance of new functional groups, and at higher temperatures - depolymerization. The formed products are adsorbed chemically, interacting with the catalytically active juvenile surface of the plastically deformed metal. These processes and phenomena impose an imprint on the picture of the mechanism of cutting or plastic deformation of metal. At the same time, the influence of the polymeric component of CRM on the machining process does not end here. It is necessary to suppose also possible diffusion of depolymerization products into the volume of contacting bodies by their interaction with electrically active structure of metal, which leads to change of energy necessary for formation of new surface. However, so far there is no experimental confirmation of this proposal.

This paragraph analyzes the results of studies on the possible change in the chemical composition of the material volume after its mechanical processing in different environments.

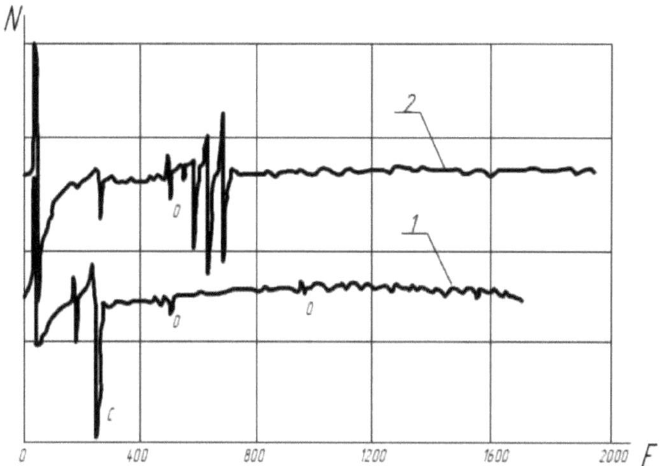

Fig. 3.6. OLE spectra taken from the surface of steel 45 after face turning in polymer-containing SOTS: 1 - on the surface; 2 - at a depth of 100nm.

OGE - electron spectroscopic analysis of samples of steel 45 and pure metals N1, Co, Si, Ni, Mo has shown that treatment in polymer-containing composition leads to the penetration of chlorine into the sample and to a considerable depth of a large amount of carbon. As for oxygen, its concentration does not exceed 12 at. % near the surface, and at a depth of 100 nm decreases to the background level (Fig. 3.6.).

To clarify the influence of mechanical processing in different SOTS on the distribution of elemental composition in the near-surface layer of the metal, the curves for different samples characterizing the changes in the concentrations of O and C at the depth from the machined surface are compared in Fig. 3.7. curves for different samples characterizing changes in concentrations of O and C at depth from the machined surface are compared. By comparing the profiles for different compositions, it is possible to trace that the composition of diffusing elements, the amount and depth of their penetration and the treated material significantly depend on the type of medium, which is used for treatment. Thus, when the samples are treated in air atmosphere, carbon is fixed only on the surface (approximately 40 at. %) and its concentration sharply decreases to the background level at a depth of approximately 10 nm (Fig. 3.8). On the surface of samples treated in emulsol ET-2 and polymer-containing medium there is a high concentration of carbon (over 80 at. %) and practically no iron (detected only at the background level).

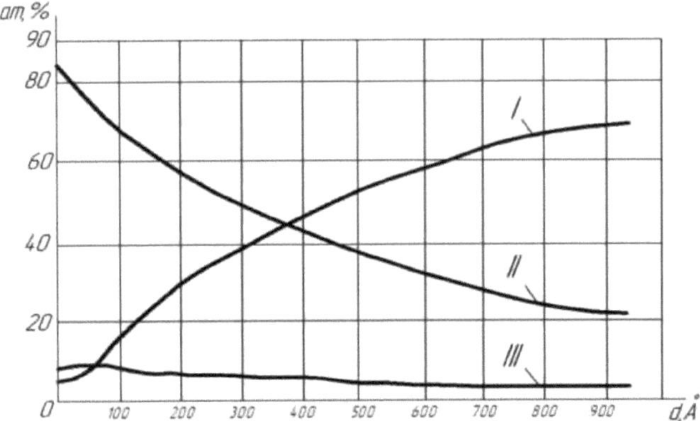

Fig. 3.7. Distribution of elements in the surface layer of 45 end turning steel in polymer-containing SOTS: 1 - iron; 2 - carbon; 3 - oxygen.

The oxygen distribution profile in contrast to carbon (Fig. 3.8) is described by curves with maxima, the position of which depends on the type of process medium used during processing (Fig. 3.9). Here it is necessary to note a strong difference in the oxygen concentration on the surface and in the near-surface layers of the samples - 8-12 at. % when processing in polymer composition and over 60-70 at. % when processing in air atmosphere.

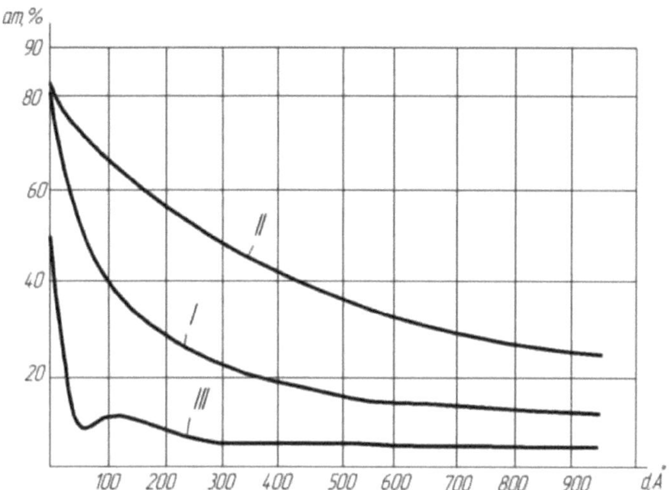

Fig. 3.8. Distribution of carbon in the surface layer of steel 45 after face turning in different media. 1 - treatment in emulsion ET-2; 2 - treatment in polymer dispersion; 3 - treatment in air

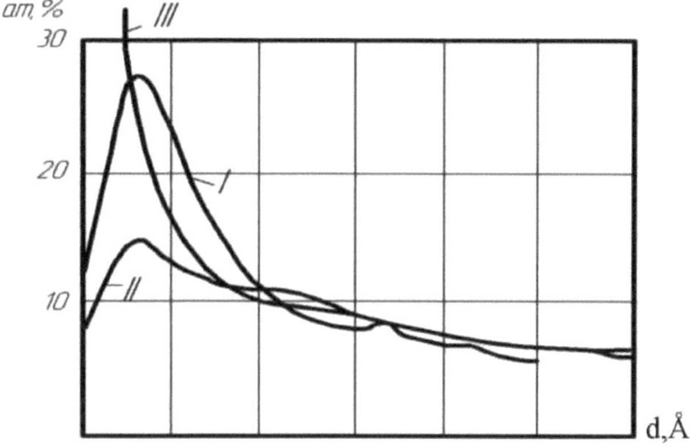

Fig. 3.9 Oxygen distribution in the surface layer of steel 45 after face turning: 1 - in emelsol ET-2; 2 - in polymer-containing SOTS; 3 - in air.

Studies have shown that when cutting Ti in water with PVC additive, C1 and C atoms, which are part of the PVC latex, are adsorbed on the surface. When processing in water, these elements are absent on the surface of titanium. It should be noted that when turning Tì both in water and in water with PVC, chemical elements diffuse deep into the metal (Fig.3.10. and 3.11.). However, for the former case, the concentration in the material volume is much lower than for the latter. For example, at a depth of 150 nm, the concentration of oxygen and carbon when Ti is treated in a polymer-containing dispersion is approximately equal to 25-30 at. %, while when it is treated in water it is only 6-7 at. %. This difference in the concentration of chemical elements at depth from the surface is explained by the catalytic effect of hydrogen on the diffusion process.

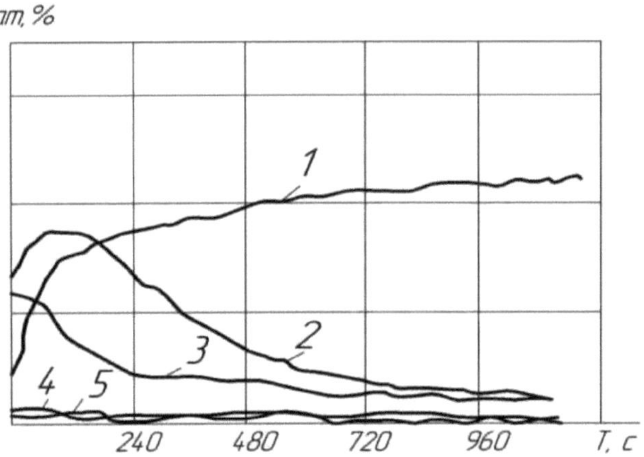

Fig. 3.10. Distribution profiles of chemical elements in the surface layer of technically pure titanium samples after end turning in water. 1 - titanium; 2 - oxygen; 3 - carbon; 4 - sulfur; 5 - chlorine.

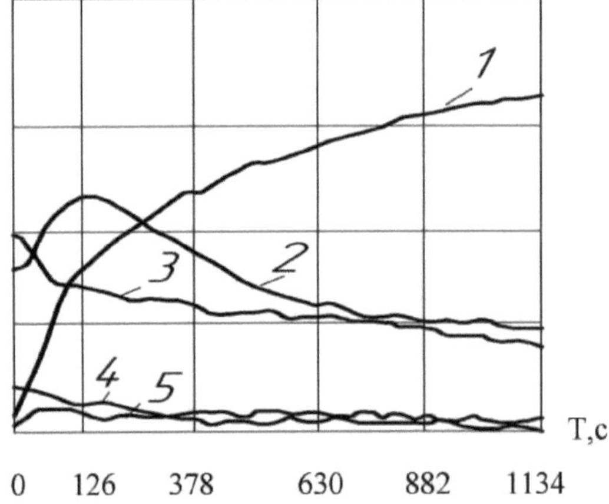

Fig. 3.11. Distribution profiles of chemical elements in the surface layer of technically pure titanium after face turning in polymer-containing SOTS. 1 - titanium; 2 - oxygen; 3 - carbon; 4 - sulfur; 5 - chlorine.

The observed high concentration of carbon in the near-surface layers of both steel and pure metals can be explained by the fact that at the point of contact between the processed material and the cutting edges of the tool, intermolecular bonding forces occur, which are as strong as interatomic bonds. This results in the transfer of an appreciable number of atoms [80-82], in this case carbon atoms, to the surface formed after the contact is broken.

The analysis of surface interactions of the cut layer with hydrogen (ionized, atomic and molecular) can be further united according to Rebinder by the general term of adsorption interactions, and there is a need to identify two issues that require a separate solution: to what extent the effect of using polymeric SOTS is associated, including with interactions in the surface layer of the workpiece, and how to quantitatively characterize the effect on these interactions of low-temperature plasma consisting of a mixture of different activities

3.2 Hydrogen adsorption

3.2.1. Adsorption on the workpiece surface

In the process of chip formation, we will be interested in the surface of the cut layer of the workpiece insofar as it interferes with the deformation process, making it difficult for dislocations to move (and nucleate and multiply). Due to the long-range action of the elastic fields of dislocations, it reveals the entire near-surface layer to a depth at least of the order of the average size of dislocation segments, or even noticeably larger. This circumstance determines the extent to which ductility turns out to be a "surface" property: precisely to the extent that the near-surface layer of the specified thickness determines a significant share of the energy and force inputs for chip formation.

On the surface of treated metals there are various types of adsorbed layers: gas, moisture, polar and non-polar molecules of organic substances. The near-surface layers of metal differ in their structure and properties from the base material. There are zones of deformed, hardened and de-strengthened material covered with oxide layers (Fig. 3.12). Depending on the method of workpiece production, its chemical composition, and the composition of the environment, chemical reactions occur between the surface layers of metal and the environment.

Of particular importance are steel oxidation reactions leading to the formation of scale on the surface of the heated product, and steel decarburization - the burning out of carbon in the surface layers.

The main oxidation reactions of steel billet are as follows:

$$Fe + O \rightarrow FeO; \; Fe + CO_2 \rightarrow FeO + CO; \; Fe + H_2O \rightarrow FeO + H_2$$

Basic decarburization reactions:

$$Fe\,y(C) + 2H_2 \rightarrow Fe\,y + CH_4; \; Fe\,y(C) + O \rightarrow Fe\,y + CO;$$

Thus, oxygen, carbon dioxide and water vapor are oxidizing media, carbon monoxide and hydrogen are reducing media, hydrogen and oxygen are decarburizing media, and methane and carbon monoxide are carburizing media.

Usually, three oxide layers are formed on the surface of the processed metal workpiece, based on iron: FeO, Fe 0_{34} , Fe O_{23} . In a humid atmosphere, the surface of iron is covered with rust. In this case, water (moisture) is not just present, but takes part in the reaction according to the scheme: $Fe + O_2 + H_2\,O \rightarrow Fe\,(OH)_3$

With the formation of hydroxide.

The role of oxide films in metal machining is twofold. On the one hand, boundary layers (films) on the rubbing surfaces of the cutting tool and the processed metal reduce friction and wear of the tool. The intensity of oxidative wear depends on the thickness and strength of the film. Its structure, composition, protective properties depend on temperature, contact pressure, time factor. On the other hand, friction of the tool with the workpiece is accompanied by activation of surface layers of the processed material, increases its ability to adsorption, diffusion, chemical reactions, to passivate the material and reduce the obstacle to the exit of dislocations on the surface, thus contribute to facilitating the flow of plastic deformation, i.e. removal of the sheared layer. In addition, plastic deformation activates the friction surface. This creates a nonequilibrium electronic state, activated state of atoms of surfaces, which contributes to the formation of knots of adhesion and the development of tool wear.

If the surface of the processed material is free of boundary films and various impurities, it does not cause stresses in the surface layer that would prevent the dislocation movement, i.e., the development of plastic deformation. As a result of decreasing dislocation energy as it approaches its free surface, the dislocation will be more and more strongly attracted to the surface until it leaves the crystal; in this case, a step with a height of one interatomic distance is formed on the surface.

If the surface of the treated material is covered with a thin impermeable layer, such as an oxide film on steel, it may be that dislocations attracted to the surface cannot leave the crystal (due to the difficulty of sliding in this layer); a noticeable surface hardening occurs.

Thus, the surface interferes in the process of cutting of the chip layer, i.e. in the deformation process, as it impedes the movement (and nucleation and multiplication) of dislocations, and this phenomenon, due to the long-range action of the elastic fields of dislocations, reveals the entire near-surface layer to a depth at least of the order of the average size of dislocation segments or even noticeably larger.

The state of the metal surface also has a great influence on the kinetics of hydrogen absorption. Oxide films in particular make it difficult for hydrogen to penetrate into the metal. For example, the oxidized film on aluminum in general prevents the absorption of hydrogen from the gaseous phase even at high pressures and temperatures. Vacuum annealing, eliminating the oxide film strongly activates the metal surface, contributing to its hydrogenation. The interaction of the deformed and then sheared metal layer with hydrogen is a complex process involving various hydrogen reactions at the surface and deep within the metal, which is discussed in later sections of the book.

Thus, the surface condition of the processed metal plays an essential role in the balance of energy spent on chip formation. Therefore, when evaluating the effectiveness of a coolant or studying the mechanism of its influence on the chip formation process and cutting tool wear, it is necessary to take into account the surface condition of the machined material.

3.2.2. Adsorption on the shear layer

The enrichment of ionized hydrogen on the surface of the metal-gas interface is due to the unbalanced interaction forces between the atoms that make up the solid, i.e., the unsaturation of bonds at the atoms lying on the surface.

The adsorption intensity of a chemical element, i.e., the intensity of the decrease in the free surface energy level, is usually the greater the stronger the chemical affinity between the metal and the interacting element.

The change in the free surface energy of metal as a result of hydrogen adsorption leads to a decrease in the energy consumption for its destruction, i.e. it contributes to the facilitation of the metal cutting process.

The heat of hydrogen chemisorption is much higher than that of physical adsorption of low-molecular substances included in conventional SOTS and approaches to the heat of chemical reactions. Thus, the heat of physical adsorption of hydrogen on iron is usually less than 10 kJ/mol, and the heat of hydrogen chemisorption is usually 10 and more times higher. In this connection, at mechanochemical treatment of steel the adsorption effect of deformation and fracture facilitation will be much greater than in COTS with no high-molecular compounds in their composition.

The analysis of the research results presented in the previous chapters allows us to believe that the effect of hydrogen, which is formed in the cutting zone as a result of pyrolytic transformations of the polymeric additive COTC, on the condition of the surface of the processed metal consists of two stages. At the first stage, adsorption of ionized hydrogen on the oxidized metal surface occurs, as a result of which the surface is cleaned from the oxide film and becomes free of various contaminants.

It should be noted that molecular hydrogen is a relatively inert gas due to the strength of the H-H bond. Whereas, hydrogen in the active state, the one formed in the cutting zone in polymer-containing SOTS, is characterized by high activity. In particular, it has high reducing properties. Thus, even at normal temperature it reduces many metal oxides by combining with oxygen.

Further, on the clean surface there is a transition of gas from physically adsorbed state to chemisorbed state. Moreover, this transition is carried out not only for ionized hydrogen, but also for atomic and even molecular hydrogen, which is pre-dissociated on the clean surface. The resulting atomic hydrogen is partially adsorbed by the surface and passes inside the metal along grain and subgrain boundaries, and also dissolves in the crystal lattice of the metal, forming a solid solution.

In addition, hydrogen can interact chemically with impurities contained in the metal. This leads to a change in the solubility of hydrogen in the crystal lattice and the formation of new phases. For example, when interacting with oxygen, water vapors 2H + O = H2O are formed; with carbon 4H + C = CH4, or with carbon in carbides 4H + Fe_3 C = 3Fe + CH4. These new gaseous products are released, as a rule, in non-continuities, mainly at grain boundaries and along slip and twinning planes [83-85].

Consequently, during mechanochemical processing of steel, shearing of the surface layer from the workpiece, i.e., the deformation process, occurs under conditions where hydrogen cleans the surface of the sheared layer (and chip surface) of various contaminants, which eliminates the obstacle for dislocation exit to the surface of the processed metal.

It is known [85,86] that the free surface does not induce stresses that would prevent dislocations from moving. As a result of energy reduction, dislocations will be more and more strongly attracted to the surface as they approach the free surface until they leave the crystal; in this case, a step with a height of one interatomic distance is formed on the surface.

If the surface is covered by a thin impermeable layer, such as an oxide film on steel, it may be that dislocations attracted to the surface cannot leave the crystal; there is a noticeable surface hardening and an increasingly high strain stress is required to "push" such dislocations through the obstacles.

Thus, at mechanochemical machining, as a result of hydrogen interaction with the surface of the cut chip, plastic deformation starts at lower stress than at machining without mechanochemical effect. Consequently, already at the first stage of the cutting process in the polymer medium, energy and power costs for chip formation are reduced in comparison with machining with the use of low-molecular component-based coolants.

When interacting with different reagents, hydrogen behaves differently, participating in covalent, ionic and metallic bonds depending on the conditions[87]. As a consequence, hydrogen forms:

а) true solutions by interaction with chromium, iron, cobalt, platinum, copper, molybdenum, aluminum, etc., hydrides of some of these metals can be obtained only preparatively;

б) hydrides with titanium, zirconium, nickel, vanadium, etc.

At the same time, it has been established that at high mechanical loads, high temperatures and deformation rates in almost all metals can form areas of mixtures of solid solutions with intermetallic compounds - hydrides, representing in most cases phases of variable chemical composition [88].

Therefore, when cutting steels alloyed with various chemical elements, hydrides may occur in them. They are located mainly along grain boundaries, sliding and twinning planes.

The embrittlement effect of hydrides is due to the brittleness of the hydride inclusions themselves (low resistance to break-off), as well as the occurrence of tensile stresses in the matrix due to the large specific volume of hydrides compared to metal.

Therefore, when analyzing the processes occurring during cutting of steel in polymer-containing *coolants, it* is necessary to take into account the possibility of *alloying chemical elements to form hydrides with hydrogen, starting from the surface of the workpiece, causing brittle fracture.*

Chapter IV

Diffusion of hydrogen into the treated metal

4.1. Some features of hydrogen diffusion

The diffusion mechanisms of embedded atoms depend significantly on temperature, and the diffusion mechanism of hydrogen differs significantly from those of other embedding impurities [89-91]. This is explained by the fact that after ionization, oxygen and nitrogen atoms have sizes comparable to those of the internodes, and hydrogen is in the lattice in the form of a proton quasi-ion screened by the electron gas. In addition, at low temperatures, a significant contribution to hydrogen diffusion is made by proton tunneling transitions from one internode to another[92].

Hydrogen in metals has an unusually large diffusion mobility. Even at temperatures below room temperature, it is large enough to redistribute in volumes comparable to the size of micrograins. At room temperature, the diffusion coefficient of hydrogen in ferrite is 1010 - 10^{11} times greater than that of carbon and nitrogen, and 10^{35} times greater than typical values of diffusion coefficients of substitutional elements.

At higher temperatures hydrogen atoms are localized in a certain internode or near it and diffusion is performed by thermally activated jumps of atoms from one internode to another. An atom can jump from one equilibrium position to another either by tunneling or as a result of acquisition of additional energy sufficient to overcome the potential barrier.

At sufficiently high temperatures, especially in o.c.c. metals, hydrogen nitrates so frequently jump from one equilibrium position to another that the average time between two consecutive jumps becomes comparable to the duration of residence of a hydrogen atom in the interstitium [92]. Under such conditions, it makes no sense to distinguish between the resting state and hopping. In this case, the diffusion of the hydrogen atom occurs in approximately the same way as in a dense gas or liquid. This mechanism of diffusion of embedding atoms is called liquid diffusion [92]. Naturally, there are no clear temperature boundaries in which one or another diffusion mechanism operates.

All the above refers to the diffusion of the introduction impurity in an ideal lattice.

The diffusion coefficient, which determines the rate of this process, is called the true or lattice diffusion coefficient. In a real metal, however, there are crystalline imperfections: grain and phase boundaries, second phase particles, pores, which have a significant influence on diffusion processes. This influence can have a twofold character. First, dislocations, grain and phase boundaries. The elongated particles of the second phase can serve as paths of preferential diffusion due to a much higher diffusion coefficient compared to the lattice of the base metal [93]. Secondly, various kinds of imperfections can be collectors or traps for the introduction impurity due to the energetically favorable location of the impurity in the imperfections themselves or in the distorted lattice near them. Depending on the binding energy of the impurity to the trap, traps are divided into irreversible traps, from which the impurity is practically not released, and reversible traps, from which the impurity is released at a sufficiently low concentration in the surrounding solid solution. The efficiency of traps increases significantly with decreasing temperature.

The mechanism of hydrogen transport into the sheared layer of the workpiece is in principle no different from the above mechanism and is largely investigated by scientists and engineers working in the field of molecular physics of solid state and physical chemistry of surface phenomena.

Until recently, based on these ideas, it was difficult to assume the possibility of hydrogen saturation of the sheared layer of steel when the cutting speed is high enough. However, this book presents experimental results that show the presence of hydrogen in the chips.

Let us now turn to the question of the mechanism of hydrogen transport into the sheared steel layer formed in the presence of polymer-containing SOTS.

It is known [94-98] that under elastic deformations of steel, its hydrogen permeability increases in proportion to the level of mechanical stresses and the degree of deformation. Under experimental cutting conditions, when the thickness of the sheared layer is several orders of magnitude less than the samples on which hydrogen permeability studies are usually carried out, and the strain rate (plastic deformation) is several orders of magnitude higher.

The transition of hydrogen atoms into the metal volume requires overcoming a high activation barrier, and the transition itself is endothermic [98-100]. In other words, the process of hydrogen penetration inside iron or steel proceeds with appreciable speed only at sufficiently high metal temperatures and high concentrations (pressures) of hydrogen [99]. The temperature of the chip at the moment of its separation from the workpiece usually does not exceed 600-700°C. At such temperatures, the diffusion rate of hydrogen into iron, as well as the partial pressure of hydrogen near the surface is too low for it to penetrate deep enough into the near-surface layers of metal.

Indeed, the characteristic depth of hydrogen penetration R can be approximated by the formula $R^2 = D1$, where D is the diffusion coefficient, t = 1\V , V is the cutting speed, 1 and t are, respectively, the characteristic size of the area of actual contact between the cutting tool and the processed material and the contact time during which the plastic deformation occurs. Typically, 1 = 1mm and V = 0.1 m/sec, which corresponds to a contact time of the order of 10^{-2} sec. The diffusion coefficient of hydrogen into iron at room temperature, according to different measurements, ranges from $2 - 10^{-13}$ to $9 - 10^{-9}$ m^2 /s [101]. Thus, even according to a highly exaggerated estimate, the hydrogen penetration depth cannot exceed 10^{-3} cm. Since hydrogen still penetrates deep into the metal under these conditions to depths of the order of mm, it remains to assume that the high rate of hydrogen transfer into the sheared layer of metal can be due to: 1) a very strong instantaneous "heating" of those degrees of freedom in the crystal lattice, which are most responsible for hydrogen transfer and 2) plastic deformation of the sheared layer caused a significant acceleration of diffusion, especially at high stresses, which can be explained by the enhanced penetration of hydrogen through some areas of d

Indeed, literature data show that the penetration of hydrogen into the metal occurs along the grain boundaries and along the body of grains, and the rate of penetration through mono- and polycrystalline metals is the same [102104]. Hydrogen in metals concentrates predominantly along grain boundaries [103]. In this, the behavior of hydrogen does not differ from the behavior of other impurities in metals.

As a result of diffusion of hydrogen into the deformed and sheared layer of metal, it can be:

1) in the inter-nodes of the crystal lattice, forming a solid solution of introduction with the metal;

2) at grain boundaries, cracks;

3) in the form of chemical compounds with impurities;

4) in the form of chemical compounds with the solvent metal, hydrides.

In a polycrystal, individual grains are separated by boundaries. One of the most important properties of the boundary is its ability to inhibit the slip process in individual grains. The presence of spatial disorientation of slip planes on different sides of the boundary prevents dislocations

to pass through it. With increasing stress applied to the polycrystal, sliding will start in grains with favorably oriented slip systems. However, the boundaries of such grains will block this sliding. The impediment of sliding at the boundaries will lead to stress concentration near the grain boundary, which strongly helps the external stress to drive the less favorably oriented slip systems.

The concept of a grain boundary as a thin amorphous layer helps to explain many grain boundary effects. At the critical combination of high temperature and low stress, grain boundary sliding and fracture can occur. Grain boundaries are the preferred sites for the accumulation of impurity atoms, and nuclei of the second phase are most often formed on them in two-phase alloys. This is undoubtedly a consequence of the strong disorder at the grain boundary, which turns it into a drain for dissolved atoms (impurities), since when such an atom hits the boundary, the lattice distortions caused by it are removed.

The disorder at the boundary makes it a preferred pathway for diffusion of dissolved atoms and hydrogen. Being a sink for dissolved atoms, the boundary and its adjacent regions may differ in chemical composition from the rest of the grain. Thus, the boundaries in polycrystals represent very effective obstacles to the sliding of dislocations, i.e., to plastic deformation.

Diffusion flows of atomic hydrogen along grain boundaries lead to interaction with atoms of other alloy phases or impurities (oxygen, nitrogen, carbon, chromium, etc.) accumulated there, "exposing" crystal surfaces along grain boundaries, removing obstacles to the movement of dislocations, i.e. facilitating the energetic process of plastic deformation.

4.2. Hydrogen penetration into steel in the process of plastic deformation

One of the currently poorly studied processes is the transfer of hydrogen into the surface layers of metals during their plastic deformation in hydrogen-containing media. In the study of such processes, the penetration of hydrogen into the metal is most often registered either directly by the content of metal hydrides in the surface layers or indirectly by gas emission from the sample during its heating.

The scope of application of the first of these methods is limited only to metals capable of forming stable compounds with hydrogen, such as titanium [105], nickel [106], etc. The second method usually determines the total hydrogen content in the sample rather than its depth distribution or energy characteristics [107].

Such studies for iron from the applied point of view are of particular interest, since hydrogenation of steels during their machining significantly affects their physical and mechanical properties and cutting parameters. However, at present, very few works devoted to this problem have been published, apparently due to significant experimental difficulties arising in the study of iron due to its high chemical activity.

The book presents the results of the study of hydrogen penetration into iron and low-carbon steel at some types of their plastic deformation in hydrogen-containing media - hydrocarbons, water and others.

To detect hydrogen, the method of temperature-programmed heating (TPH) of a sample placed in a vacuumized volume with simultaneous mass spectrometric registration of the released hydrogen (see, for example, the method of temperature-programmed desorption) was applied [73]. The experimental curves obtained in this process represent the dependences of the mass spectrometer signal at a given mass (proportional in our case to the hydrogen or deuterium release rate) on the sample temperature, which, as a rule, rose at a rate of 0.5 K/sec.

Usually, the curve (TPN spectrum) has one or more bell-shaped peaks with maxima corresponding to the hydrogen yield temperature of a certain "kind". The position of the maxima on the curve, as well as the shape of the peaks, reflect a complex process occurring under conditions of linear temperature rise, the main components of which are the diffusion of hydrogen atoms in the metal, their escape and recombination on the surface with the release of molecular hydrogen. Obviously, the distribution of hydrogen over the depth of the sample before the onset of temperature rise affects the shape and position of the peaks.

Two types of plastic deformation observed in compression and cutting, respectively, were studied in this work. The objects of study were armco iron and steel 20.

4.3. Hydrogen diffusion in compression

The fact of hydrogen penetration inside the metal was established in the experiment with compression of iron armco in the medium D_2 0. A cubic shaped sample (1=3mm) was placed between plane-parallel surfaces inside a D20 drop. After applying a load of 20 tons, the specimen was compressed to a thickness of 0.5 mm. A 30 μm thick layer was then sanded off both sides of the sample to remove surface deuterated compounds. The use of labeled water is due to the need to exclude the registration of "extraneous" hydrogen present in the original sample or introduced to its surface as a result of adsorption of atmospheric moisture during grinding. In the TPN spectrum of the deformed sample after removal of the surface layer, a deuterium emission peak with a maximum around 540K was detected (Fig. 4.1), which proves its penetration into iron.

4.4. Diffusion of hydrogen into the machined metal in the cutting process

Most of the data were obtained in experiments with cutting (drilling), which, as it is known, is accompanied by significant plasticity.

by deformation of the material to be machined, and at high speed. Both steel 20 and armco iron samples were drilled. Cutting parameters: RPM - 450 rpm, feed rate - 5 mm/min, drill bit - steel P8M5. Drilling was carried out in the following media: H20, D20 atmospheric air, dry nitrogen, heptane, vaseline oil, ethanol, ethanol-water mixtures, vaseline oil-water and polyethylene-water emulsions. In addition, a sample of steel pre-saturated with hydrogen by electrolytic method was drilled. In the drilling experiments the object of TPN was chips, and the same weight of chips (50 mg) was always taken. All drilling experiments were carried out with steel and iron samples pre-annealed at a temperature of 700^0 K for 10 hours. This is due to the fact that residual metallurgical hydrogen may be present in the initial metal blanks. Thus, residual hydrogen was detected in bars of steel 45, used by us as initial billets, and the peak maximum in the TPN spectrum of unannealed steel was at 500^0 K, and the peak intensity corresponded to the usual concentration of metallurgical hydrogen $(1\text{-}5)\cdot 10^{-6}$.

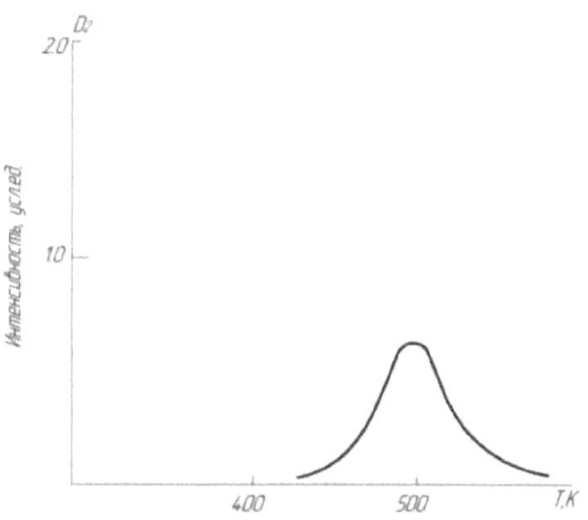

Fig. 4.1. Mass spectrometer signal dependence on sample temperature during plastic deformation.

Table 4.1. summarizes the results of determining the position of the maxima (Tmax.) on the TPA curves. as well as the dwell times(1) of the chip samples from the start of drilling (steel 20) to the time of measuring the TPA spectrum at room temperature for drilling in different media.

Table 4.1.

№	Wednesday	T(hour)	T° max (K)
1	Air	0.5	480
2	Dry nitrogen	0.5	-
3	H2O	0.5	425
4	H2O	2	445
5	H2O	23	493
6	H2O	300	-
7	H2O	1.0	490
8	D2O	0.5	425
9	D_2 0 (coarse fraction)	7	480
10	D_2 0 (fine fraction)	7	490
11	D_2 0 (coarse fraction)	50	505
12	D_2 0 (fine fraction)	50	505
13	electrolysis	0.5	475
14	electrolysis	29	480
15	ethanol	0.5	480
16	ethanol + water	0.5	443
17	n-heptane	0.5	455,505
18	petroleum jelly	0.5	455,505
19	oleic acid	0.5	500

20	petroleum jelly + water	0.5	425,480

Note: 0.5 hours is the minimum time required to prepare the experiment for measuring the TPN spectrum.

Fig. 4.2 shows in comparison the TPN spectra of chips obtained by drilling in water, measured respectively after 0.5, 2, 23 and 300 hours (curves 1, 2, 3, 4, respectively) after drilling. The peak in the TPN spectrum of freshly prepared chips is observed at the lowest temperature and has the highest integrated intensity. The intensity value indicates that during drilling in water, a total amount of hydrogen $(1\text{-}10)\cdot 10^{-6}$, i.e. comparable to the metallurgical hydrogen content, penetrates into the chips.

As the exposure time of the chips to the spectrum increases, the peak shifts to the high-temperature region with a simultaneous drop in intensity. At sufficiently long holding times (several days) the intensity drops to zero. In the case of increasing the holding temperature of chips before taking the spectrum of TPN, the process of hydrogen redistribution is accelerated, and the shift of the peak to the high-temperature region is observed for a shorter time.

In order to exclude the possibility of manifestation of "extraneous" hydrogen in the spectrum of TPN, experiments with drilling in D_2O were also carried out. Their results are absolutely identical to those for H_2O (Tab. 4.1.).

In order to determine the possible scatter in the position of maxima on the TPA curve due to the non-reproducibility of chip size from experiment to experiment, chips were fractionated by size on sieves (0.3 mm) with subsequent TPA spectra acquisition. From the results in the table, it follows that the particle size obtained in the drilling experiments practically does not affect the position of the peaks in the TPN spectra of the chips aged for 7 and 50 hours.

To evaluate the depth of hydrogen penetration into the billet after the first drill pass in D_2O environment, the hole was drilled in air with a larger diameter drill bit. When the drill diameter of the second pass was increased by 1 mm, the TPN spectrum of the workpiece showed a D peak with an intensity of 10% of the initial one, which qualitatively indicates the depth of deuterium penetration into the metal of the order of a millimeter.

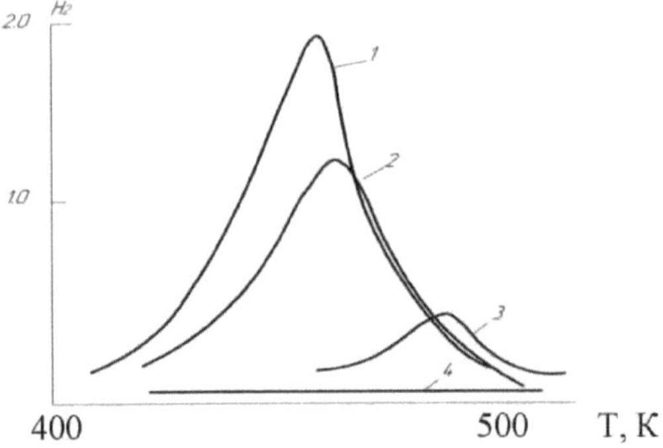

Fig.4.2 Dependence of mass spectrometer signal on sample temperature after drilling in H20. 1 - exposure time - 0.5 hours; 2 - 2.0 hours; 3 - 23.0 hours; 4 - 300 hours.

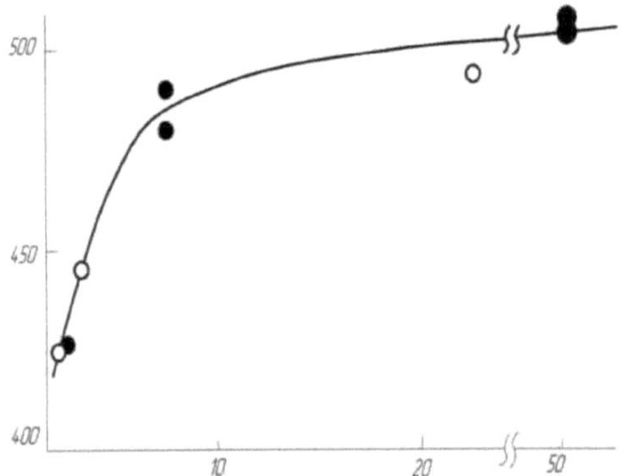

Fig. 4.3 Dependence of the temperature corresponding to the maximum hydrogen yield from the sample on the time of its heating after drilling: - D_2 0; O - N_2 0.

The data on the position of the maximum on the TPN curve for chips obtained in water depending on the time of their preliminary curing is conveniently presented in graphical form (Fig. 4.3). The figure shows that the dependence is asymptotic, with the maximum temperature reaching its limit at curing times of about 6 hours.

These results can be easily interpreted if we assume that, as in the above experiments with compression, during cutting the plastic deformation of metal is accompanied by hydrogen transport into the surface layer. Indeed, penetrating inside the chip during cutting, hydrogen concentrates mainly near the surface. When the chip is held at room temperature, hydrogen in it is gradually redistributed: by diffusion from the surface layer into the material, which leads to an equalization of concentration over the depth and by hydrogen transport to the surface with subsequent recombination and formation of molecular hydrogen-containing surface compounds and removal into the gas phase.

Redistribution of hydrogen, inside the metal with simultaneous decrease of its total content in the sample just leads to the observed high-temperature shift of the peak in the spectrum of TPN and a drop in the integral intensity as the chips are aged before TPN.

In order to validate the proposed interpretation, a drilling experiment was carried out on electrolytically prehydrogenated steel.

Electrolysis was carried out in 0.1% solution of $H_2\ SO_4$ with current density of 300 mA/cm for 2 hours. To exclude the ingress of "extraneous" hydrogen, drilling was carried out under dry conditions (nitrogen), with the portion of chips obtained at 1-2 mm of surface layer separated and not used further.

The data obtained were compared with the results of drilling samples in water with the addition (1%) of fine polyethylene powder.

The results are presented in Figure 4.4 and Table 4.1.

The measurement was performed after 0.5 hours (curve - 1 and, and 29 hours (curve 2).

First of all. it can be seen that electrolytic saturation leads to the "pumping" of significantly larger amounts of hydrogen into the sample. The hydrogen content in the same amount of chips when drilling in water of the annealed sample and drilling in dry nitrogen of the electrolytically hydrogenated sample correlate approximately as 1/5. In addition, the position of the peak maximum in the electrolysis experiment corresponds to the position of the peak maximum for chips obtained by drilling in $H_2\ O$ and aged for at least 10 hours before spectrum acquisition.

This was to be expected, since during the preliminary exposure of the chips obtained by drilling in H_2O for several hours, hydrogen is redistributed more uniformly throughout its volume, penetrating into deeper layers, which leads to the high-temperature shift of the peak. In the chips obtained by drilling the electrolytically hydrogenated sample, hydrogen is distributed more uniformly over the volume from the very beginning.

Similar results (Fig.4.5 curve 3) were obtained when drilling in water with polyethylene addition, which is an indirect evidence of diffusion of polymer thermomechanodegradation products (hydrogen) into the metal volume.

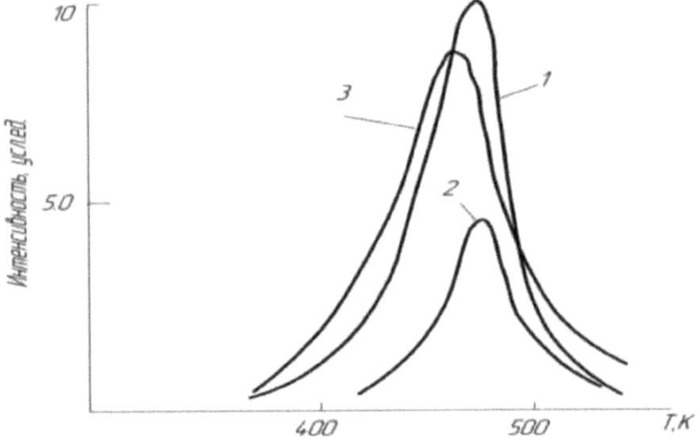

Fig. 4.4. Mass spectrometer signal dependence on sample temperature after drilling in air 1,2 - electrolytically hydrogenated samples, 3 - samples coated with polyethylene.

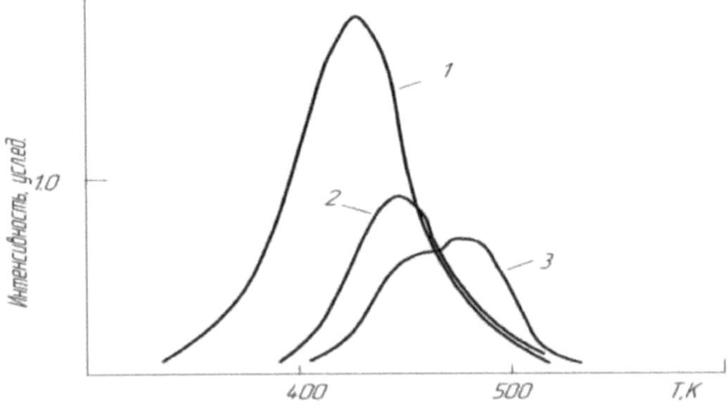

Fig. 4.5. Mass spectrometer signal dependence on sample temperature after drilling: 1 - in water; 2 - in ethanol; 3 - water + ethanol in the ratio 1:1

Let us now turn to the question of the mechanism of hydrogen transport. In general terms, it can be presented in the following form. In the course of plastic deformation, both in compression and cutting, a chemically pure (juvenile) metal surface is continuously formed, with which the medium molecules react. It is known that the reaction of such hydrogen-containing molecules as water, limiting and unsaturated hydrocarbons, alcohols and ketones with pure iron is accompanied by their dehydrogenation and the appearance of hydrogen atoms chemically bound to it on the surface [103,109]. It is also known that the transition of hydrogen atoms into the volume requires overcoming a high activation barrier - and the transition itself is endothermic. In other words, the process of hydrogen penetration inside iron proceeds with appreciable rates only at sufficiently high metal temperatures and significant concentrations (pressures) of hydrogen. The chip temperature at the moment of its separation from the workpiece usually does not exceed 600 - 700C. At such temperatures, the diffusion rate of hydrogen in iron, as well as the partial pressure of hydrogen near the surface is too low for it to penetrate deep enough into the near-surface layers of metal.

The spectrum of TPN of chips obtained by drilling in ethanol after soaking for 0.5 h is shown in Fig. 4.5. It can be seen that the position of the peak maximum is significantly shifted to the high-temperature region (curve 3) compared to the corresponding value for water (curve 1). As for the peak intensities in the case of alcohol and water, although they are generally of the same order, the intensity is always slightly lower for alcohol. The same figure shows the TPN curve for drilling in a 1:1 ethanol-water solution (curve 2) and, for comparison, the TPN spectrum for drilling in water. It can be seen that the peak maximum for drilling in solution is in an intermediate position. Looking ahead we note that drilling in organic media, as experience shows, always leads to a high-temperature shift of the maximum compared to water. This can be explained as follows.

When drilling in water, the chips cool down significantly faster than in the case of other solvents due to the special thermophysical properties of water. Thus, in the case of organic media, the chip is heated for a longer time and during this time the hydrogen has time to distribute more evenly over its volume than in the case of water. This explanation is supported by the fact that when drilling in air atmosphere (containing water vapor), the peak emerges at a higher temperature than when drilling in water (see table). Although in both cases the hydrogen-containing substance (hydrogen source) is water, the chips cool longer in air than in water and therefore the hydrogen has time to diffuse deep into the sample. Water-alcohol mixtures have intermediate thermophysical properties, which leads to the intermediate position of the maximum on the TPN curve.

Fig. 4.6. shows the spectrum of TPN of chips obtained by drilling in heptane (curve 1), vaseline oil (curve 2) and oleic acid (curve 3). After soaking for 0.5 hours, the spectra for the first two substances are almost identical. They have two maxima (or a maximum and a shoulder). The high-temperature maximum, in all cases more intense and most likely refers to hydrogen diffusing into the surface layers during the cutting process. This conclusion is further supported by the fact that the corresponding maximum for oleic acid is at the same temperature. As for the low-temperature shoulder, its origin is not quite clear. It is possible that it refers to the desorption of hydrogen from surface compounds, but this explanation needs further verification. As in the case of alcohol, the shift of the maximum to the high-temperature region is explained by the worse conditions of heat transfer from the chip to the medium for hydrocarbons compared to water.

This experiment was caused by the observed in practice sharp increase in the efficiency of cooling lubricants when using aqueous emulsions or suspensions of polymers, in particular polyethylene [105].

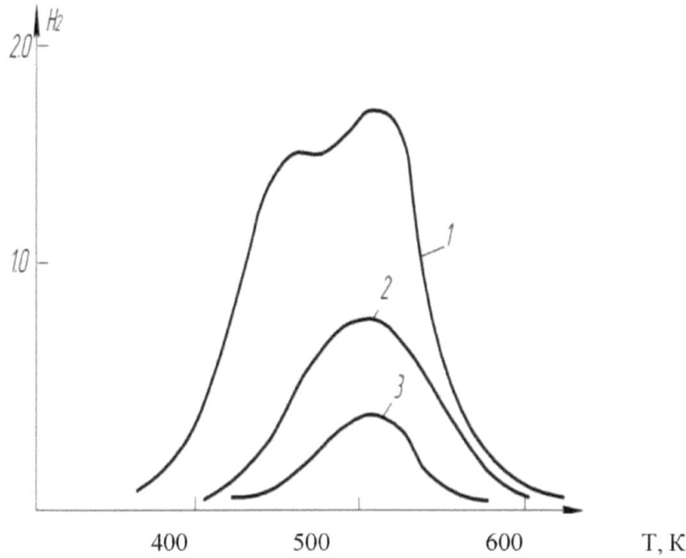

Fig. 4.6. Mass spectrometer signal dependence on sample temperature after drilling: 1 - H-thentane; 2 - vaseline oil; 3 - oleic acid.

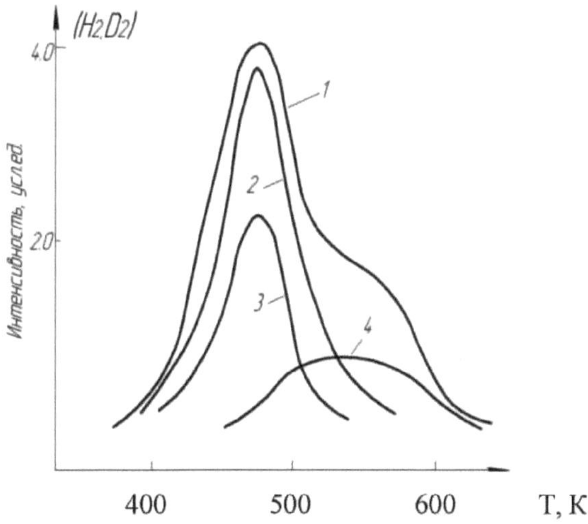

Figure 4.7. Mass spectrometer signal dependence on the sample temperature after drilling: 1 - H2/BM + H_2O; 2 - D_2/BM + D_2O; 3 - H_2/H_2O; 4 - H2/BM + D_2O.

Fig. 4.7 shows the TPN curves of chips obtained by drilling in emulsions of vaseline oil in H_2O and D2O. The spectrum for chips obtained by drilling in water is also shown there for comparison (the mass spectrometer was tuned to register both H_2 and D_2). The figure shows, first, that of the two peaks observed in the emulsion spectrum, the low-temperature one refers to hydrogen absorbed by the metal from water, while the high-temperature one refers to hydrogen from organics. This is evidenced by the fact that when drilling in the emulsion of vaseline oil in D_2O the high-temperature peak is represented mainly by H_2 , and the low-temperature peak by D_2. The presence of two peaks in the spectrum simultaneously indicates that the chip material additively includes regions of two types. The first type region arises from cutting in water medium and the second type region in hydrocarbon. Secondly, the figure demonstrates that when an emulsifier is added to water, the peak corresponding to hydrogen from water increases dramatically. In other words, the emulsion promotes the penetration of more hydrogen, from water. It is possible that this fact is directly related to the greater effectiveness of emulsions as lubricating and cooling fluids.

Thus, the conducted researches prove that high efficiency of polymer-based SOTS is mainly connected with formation of hydrogen in active form in the zone of mechanical treatment, which facilitates the process of plastic deformation of metal.

The high rate of hydrogen penetration into the zone of metal pre-destruction can be related to the high concentration of hydrogen protons adsorbed on the metal surface, which is confirmed by the following estimation.

A gas can be adsorbed on a metal surface that is one or two molecules thick. This is because, as Langmuir points out, the area of action of the surface forces responsible for the adsorption phenomenon is of the order of 10^{-8} cm. This means that the surface force region is usually smaller than the diameter of most gas molecules. Consequently, in the case of true adsorption, the thickness of the adsorbed layer cannot be more than one molecule, because as soon as a layer one molecule thick is formed on the surface, the surface forces are chemically saturated. Since the diameter of a hydrogen proton adsorbed by the surface is approximately 10-5Â, it follows that about 100 thousand layers of a hydrogen proton may appear in the area of action of surface forces, i.e. in this case the concentration factor will influence the rate of diffusion processes.

Chapter V

Diffusion of carbon into the cutting tool blade

The durability of a cutting tool significantly depends on the properties of its surface layer: hardness, tendency and adhesion, diffusion mobility of chemical elements, phase composition and microstructure. These properties change during operation under the influence of temperature and force factors as a result of interaction with the material being machined. Electronographic and micro-X-ray spectral analysis of the surface layer of high-speed steel, carried out after tool operation, showed that such elements as chromium, tungsten, cobalt diffuse from the cutter material. Carbon, which has a high diffusion coefficient, diffuses effectively. Iron[110] diffuses actively from the material being machined. The influence of diffusion on durability can be manifested not only in direct mass transfer, but also in other ways, for example, through weakening of the tool matrix due to diffusion migration of hardening chemical elements: carbon, tungsten, chromium and others. This leads to intensification of other processes, in particular, adhesion processes observed microscopically in the form of detachment of individual grains and blocks of tool material.

Changing the composition of coolants can lead to diffusion saturation of the surface layer of the cutting tool, with a local temperature increase on the surface causing process intensification [111]. Conventional coolants cause surface oxidation. Modified composition of the process medium by appropriate chemical compounds: aqueous ammonia solution, radon compounds, phosphate layers and others - can lead to diffusion saturation with nitrogen, carbon, formation in the surface layer of phosphate, sulfate films and other compounds and phases that reduce friction and contribute to increased durability of the cutting tool.

Media containing polymer components are promising in terms of providing high durability of cutting tools. High tool durability in such technological media is apparently due to some features introduced by polymers into the physical and chemical processes occurring on the contact surfaces during machining [73].

The peculiarity of the mechanism of activating action of polymers at elevated temperatures existing in the deformation and fracture zone during cutting is the possible intensification of the processes of diffusion of atom-active products of polymer degradation (C, H, N and other chemical elements) into the surface layers of the processed material and the cutting edge of the tool under the action of hydrogen [80].

In this connection, it is of interest to study the surface layer of the tool after working in polymeric technological medium in order to determine the influence of the latter on the process of saturation of cutting edges of the tool with carbon. Polyethylene was used as a polymer additive to SOTS, which gives only active elements of hydrogen and carbon during thermodestruction. Therefore, the increase in wear resistance of cutting tools should be associated not only with the reduction of energy and power costs of the machining process, but also possibly due to the diffusion of carbon into the cutting edge.

Drill bits made of P6M5 steel after cutting in different coolants were investigated. By the method of comparison of X-ray radiographs of the output sample and after machining in a certain medium, the structural changes in its surface layer, the possibility of formation of new phases, as a result of interaction of the surface of the cutting edge of the tool with the molecules of the medium, which is formed in the cutting zone due to the thermal destruction of the polymeric component of the coolant were studied.

The X-ray diffraction pattern of the initial structure of the drills that were worked in COTS without polymer (Fig.5.1) has characteristic lines of α- and β-iron and Fe carbide W_{33} C.

X-ray diffraction lines Fe W_{33} C and lines of residual austenite *(j)* appear with increasing cutting modes and time of operation of the drill. The presence of residual austenite lines allows us to conclude about the diffusion of carbon into the cutting edge of the tool.

Apparently, high temperature in the cutting zone not only activates carbon diffusion, but also positively affects the course of chemical reactions with the formation of a new modification of a-iron (martensite), which has a volume-centered tetragonal lattice. Significant diffusion of carbon into the tool blade and changes in its structure are also evidenced by the results of microstructural analysis. Using the dependence between the doublet lines on the c/a ratio of the tetragonal lattice, the amount of carbon in the martensite was determined. For example, using the data of Fig. 5.12, the amount of carbon in the martensite was calculated when the drill bit was operated for 150 hours in the COTS without polymer 2 minutes from operation in the COTS with polymer.

If in the first case the carbon content in the tool blade amounted to 0.85%, the amount of carbon in the tool blade increased 2 times during 2 minutes of its operation when working in SOTS with polymer.

Thus, during the operation of the tool, a new modification of α-iron - martensite - is formed in its blade, which is characterized by significant wear resistance.

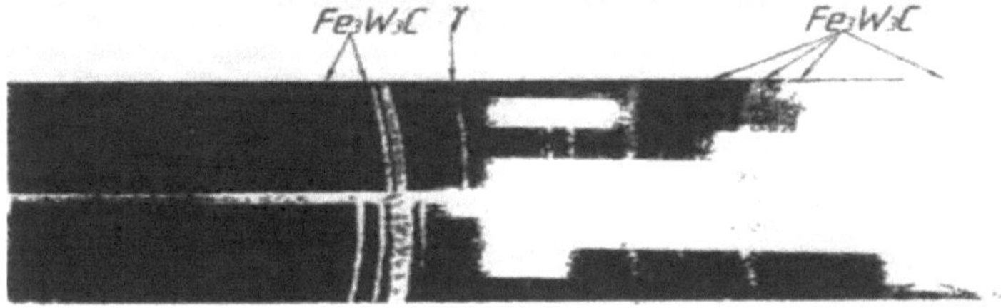

Fig. 5.1. Radiograph of P6M5 drill bit after working in polymer-free (1) and in polymer-added (2) coolants.

Qualitative assessment of diffusion activity of the products of thermodegradation of polymer component of CRM was carried out by the method of physical modeling [97], the similarity of the following parameters is observed: the sample heating temperature corresponds to the maximum temperatures in the process of cutting; the chemical composition of the bath in which the samples are heated is identical to the composition of the applied technological medium; the samples are made of appropriate tool and machining materials; the dwell time of the samples is tentatively assumed to be equal to Samples are immersed in a wave with technological medium and heated by electric current, the heating temperature is controlled by a thermocouple attached to the sample.

After making specimen slides, the main diffusion effect of this technological medium is evaluated.

Drills of 4 mm diameter made of P6M5 material were investigated in the model composition: 5% polyethylene solution in MS-20 industrial oil. The samples were heated at temperatures of 650°C, 750°C, 850°C for 20, 40 and 60 seconds.

The dependences presented in Fig.5.2, 5.3 and Table 5.1 indicate that even at insignificant contact time with the model medium at operating temperature there is a diffusion of carbon into the sample, which is carried out at a higher rate.

From the presented research results, the high rate of carbon saturation of the tool draws attention. Apparently, this can be attributed to the fact that the presence of hydrogen in the degradation products influences the process of tool saturation with carbon, providing a significant reduction in the duration of the process [112]. Usually, this is explained by the increase in the diffusive mobility of carbon in the Nernst layer and "double electric layer" under the influence of hydrogen, which increases the diffusive mobility of carbon in the near-surface layer of the tool [112].

Table 5.1.

Diffusion layer depth and calculated carbon saturation rate of the sample

№	Processing temperature, °C	Processing time, sec.	Layer depth, mm	Saturation rate, mm/min.
1	850	20	0,040	0,120
		40	0,065	0,097
		60	0,080	0,085
2	950	3	0,025	0,500

Hμ, кг\мм^2

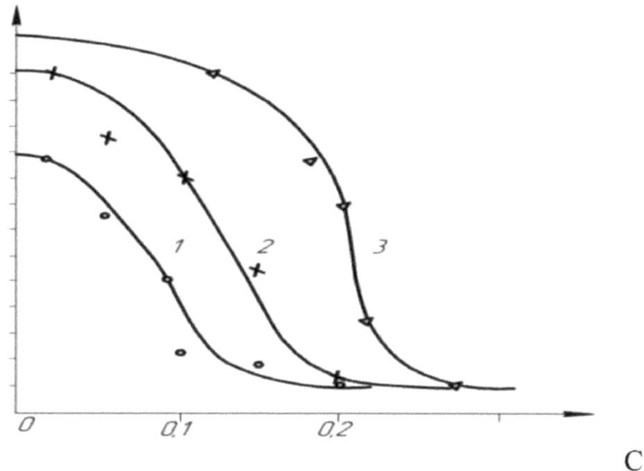

C,%

Fig. 5.3 Distribution of microhardness (Hμ) along the depth of the specimen at treatment temperature 850-C and duration of 20 (1); 40 (2) and 60 (3) sec, for 5% concentration of polyethylene in industrial oil.

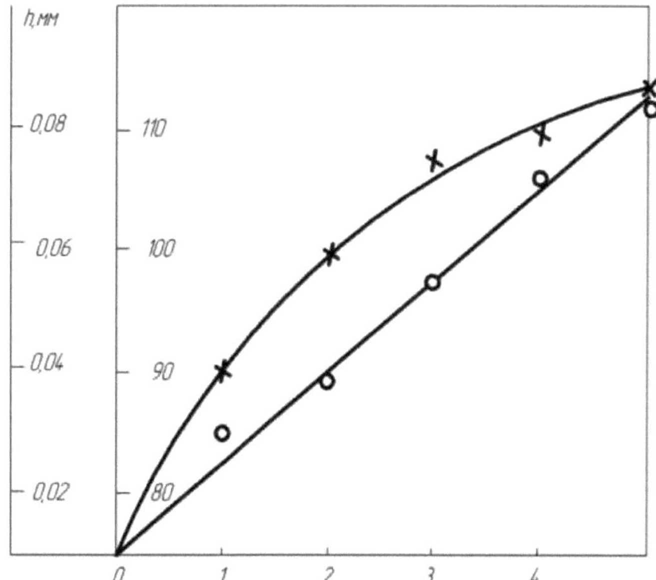

Fig. 5.4. Dependence of the saturation layer depth h (1) and microhardness on the tool surface Nm (2) on the polyethylene concentration (C) in SOTS at 850°C and dwell time of 20 sec.

The experience of using polymer-containing media at machine-building enterprises has shown that during operation a carbon layer accumulates on the surface of the cutting edge of the tool. X-ray diffraction studies have shown that this layer is amorphized carbon, which is used as a lubricant in friction units operating under high contact loads. Therefore, high wear resistance of cutting tools operating in polymer-containing SOTS should be associated not only with the reduction of energy and force expenditures for the process of plastic deformation of metal, formation of hard carbide phase in the tool blade, but also with the emergence of a layer of amorphized carbon, which is a good lubricant separating the rubbing surfaces.

Thus, diffusion of carbon, a product of pyrolysis of the polymeric component of COTS, into the cutting edge of the tool has been established. The depth and speed of carbon saturation have been estimated. It is shown that a significant reduction in the process duration is due to the influence of hydrogen.

It is also established that the increase in wear resistance of cutting tools when working with polymer-containing SOTS is the result of not only facilitating the process of deformation and destruction of material in front of the tool blade, but also due to the diffusion of carbon into the cutting edge of the tool and the formation of a new modification of α-iron - martensite.

Literature

1. P.E.Bertach Rubb. World. - №3, 1961- p. 73

2. Chemical reactions of polymers. Coll. - M.: Mir, 1967. - vol. 2 - p. 535

3. M. Lethor. Investigation of the mechanism of some chain reactions // Chemical kinetics and chain reactions. - Moscow: Nauka, 1965 - p.265.

4. P.Le Geoff, M.J. Le Tort // Chem, Phys. - 1954, №3 - 27 p.

5. N.Grassia, H.W.Melville // Pros.Roy.Soc. - 1949. - A 129, No.1 - p.14

6. N.Grassia, H.W.Melville // Chem.Abatz. - 1950, - № 44, p.71

7. A.I.Soshko, V.A.Soshko. Lubricating and cooling technological means in mechanical processing of metals. - Kherson.: Oldi-Plus.- 2008, Part 1, II. p.388

8. V.I.Likhtman, E.D.Shchukin, P.A.Rebinder. Physico-chemical mechanics of metals. Izd. of the USSR Academy of Sciences, 1962, p.217.

9. G.V.Karpenko, R.I.Kripiakevich. Influence of hydrogen on structure and properties of steels. M.: Metallurgizdat, 1962. p.198.

10. П. Cottrell. Hydrogen embrittlement of metals. Moscow: Metallurgizdat, 1963, p. 117

11. B.A. Kalachev. Hydrogen embrittlement of non-ferrous metals. Moscow: Metallurgizdat, 1966, p. 256.

12. L.S.Moroz, B.B.Chechulin. Hydrogen embrittlement of metals. Moscow: Metallurgy, 1967, p. 256.

13. Y.M.Potak. Brittle Fractures of Steel. Moscow: Oborongiz, 1955, p.389.

14. A.K.Litvin, V.I.Tkachev, // FCHMM, Vol. 12, No. 2, 1976, p. 23

15. G.V.Karpenko, A.K.Litvin, A.I.Soshko, : FCHMM, No. 4, - 1973, p. 37

16. G.V.Karpenko, V.I.Tkachev, : DAN USSR, 1974, - 214, - No.1, - p.317

17. A.I.Soshko, V.A.Soshko, // Vestnik of Engine Building, - Zaporozhye, - Izd. Mashinostroenie, - № 3, 2004, p.77.

18. K.R. Kapp. Uspekhi physics of metals. M.- Metallurgizdat, vol.2, - 1958, p.177.

19. C.J.Smithells Gases and Metal, Chapnan a.Hall, L., 1937, - p.217

20. V.I.Mikheeva, Hydrides of transition metals, Izd. of the USSR Academy of Sciences, - 1968, p.277

21. A.E.Vol. Structure and properties of double metallic systems. - M.: Fizmatizdat,

22. L.I.Sokolskaya. Gases in light metals, - M.: Metallurgizdat, - 1959, - p. 97

23. A.Dyakonov, A.Samarin. Izv. ANSSR, - OTN, - No. 9, 1945, p. 813

24. R.A.Andrievsky, Y.S.Umansky. Phases of implementation, M.: Nauka, - 1977, - p. 240

25. R.A.Andrievskiy, V.P.Yanchur, // FHOM, : 1975, №3, - p.154

26. V.P.Yanchur, R.A.Andrievsky, // FHOM, : 1973, No.1, - p.124

27. B.M. Trepnell. Chemisorption. Moscow: Izd. Il.- 1958, - p.328.

28. E. Fromm, E. Gebhardt. Gases and Carbon in Metals: per. from German, M.: Metallurgy - 1980, p. 711.

29. K. Smitells. Gases and Metals, M.: Metallurgizdat, - 1940, p. 228

30. G.C.Bond. Catalysis by Metals. New York: Academie Press, -1962, - p.519

31. O.V.Krylov, V.D.Kiselev // Adsorption and catalysis on transition metals and oxides. M.: Khimiya, - 1981. - c.288

32. Hydrogen in metals. vol. 1. Basic properties / Edited by G. Alefeld: M.: Mir, - 1981.- p. 475

33. S.Z.Bokshtein. Structure and structure of metal alloys. M.: Metalurgy. - 1971 - c. 496

34. G.V. Karpenko, A.K. Litvin, V.I. Tkachev, A.I. Soshko, FCHMM, 1973. №4, c.76-84

35. P.A. Rebinder, Physico-chemical mechanics, "Znanie", 1958. p.280

36. P.A. Rebinder, E.D. Shchukin, UFN, 1972, No.1.p.35-42

37. G.V. Karpenko, Vpllyv vodnyu na mekhashchshih vlast steel Vid. AN URSR, 1960. p.485.

38. G.V. Karpenko, Strength of Steel in Corrosive Environment, Mashgiz, 1963. pp. 396.

39. Retch, Phil. Mg., 1956, No. 1,p.331 -335.

40. A. Tetelmen, Sb. "Destruction of solid bodies", "metallurgy" 1967. - c.217

41. Д. Horiuchi, T. Toya, Sb. "Surface properties of solids", Mir, 1972. pp. 334-397.

42. A. Gowthmy, R. Cunningham, Coll. "Catalysis. Investigation of the surface of catalysts", IL, 1960.

43. C. Ponets, Coll. "Problems of kinetics and catalysis". IL. 1970. c.77-133

44. H. Otami, Zh-l "Tetsu po Haganz", 1974, pp.60-78, No.1 (Translation No. Ts-352333, All-Union Translation Center, M. 1973).

45. V.I. Likhtman, E.D. Shchukin, P.A. Rebinder, Physico-chemical mechanics of metals. Izd. of the Academy of Sciences of the USSR, 1962.

46. Y.V. Goryunov, N.V. Pertsov, B.D. Summ, The Rebinder Effect, "Nauka", 1966. p.376

47. E.D. Shchukin, Sb. "Sensitivity of mechanical properties to the action of the medium", "Mir", 1969.p.221-305

48. G. Sandos, Metallurgical Transaction, 1972, pp.33-38, No.5 (Translation Ts-8964, All-Union Translation Center, M., 1973).

49. I.Brown, W.M.Baldwin, J. Metals, 1954, No.6, pp.298-321

50. V.I. Tkachev, A.K. Litvin, V.A. Tetersky. A.I. Soshko, Problems of Strength, 1972, No.12.

51. Б. Trepnell, Chemisorption, Il, 1958.p.257.

52. N.S. Rogers, Asa Metallurg, 1956, p.44-60. №2.

53. I. Galland, R. AZOU, R. Bastien. C.R. Asad. SC, Paris, 1969, 268, pp. 27 - 32.

54. S.I. Mikitishin, I.I. Vasilenko, Collected Works. "Influence of working media on the properties of materials", vol. 3, "naukova dumka". 1964.

55. V.A. Teterskii, A.I. Soshko, A.N. Tynnyi, G.V. Karpenko, Coll. "Influence of working media on the properties of materials", vol. 3, "Naukova Dumka", 1964.

56. M.M. Shved, N.Y. Yaremchenko, V.S. Fedchenko, FCHMM, 1970, No. 3.

57. G.V. Karpenko. I.I. Vasilenko. M.G. Khitarishvili. V.S. Fedchenko. DAN SSSR, 1969, 185. NO. 5.

58. G.G. Napsosk, HHJohnson, Trans. AlME, 1966, 236, no. 4.

59. O.N. Romaniv, Y.V. Zima, G.N. Nikiforchin, N.L. Kuklyak, FCHMM, 1975, no. 2.

60. M.M. Shved, N.Y. Yaremchenko, FCHMM, Vol. 9, No. 3, 1973.

61. A.S.Ietelman, Robertson, A1ME, 1962, 224, p.775.

62. G. SNAii^, L. Mogeai, Arch. Metallkunde, 1948, 9, p.308.

63. Besnard, Annales de Chimie, 1961, 6, (3\4), p.245-283.

64. G.V. Karpenko, V.I. Tkachev. A.K. Litvin, DAN USSR, 1974, 214, No. 1.

65. V.I. Tkachev. Influence of working media FHMM, 1971, No. 3.

66. Copyright Certificate No. 338562.

67. A.B. Kuslitsky. Properties of metals. FHMM, 1966, NO. 3.

68. I.P. Pistun. Working environments. FHMM, 1973, NO. 6.

69. Y.M. Potak. High-strength steels, "Metallurgy", 1972.p.340

70. O.N. Romaniv, N.L. Kuklyak. Working media and metal cutting. FKHMM, 1968, NO. 2.

71. L.A. Plavich, N.P. Zhuk. M.L. Bernstein. Friction in machine parts. FCHMM, 1970, NO. 3.

72. G.V. Karpenko, V.I. Tkachev. A.K. Litvin, "Theses of Reports of the VI All-Union Conference on Physical and Chemical Mechanics of Materials", 1974, pp.80-

82.

73. A.I. Soshko, V.A. Soshko Lubricating and cooling technological means, 4.II, Kherson - 2008, - with. 388

74. A.A. Silin et al, DAN USSR, 1971,200, No. 1.

75. D.N. Garkunov, I.V. Kragelskii, A.A. Polyakov, Selective transfer in friction nodes, "Transport", 1969. pp. 380

76. N.N.Zorev. Questions of mechanics of metal cutting process. - M.: Mashgiz,1956,p.277.

77. Lubricating and cooling technological means for metal cutting / Edited by S.G.Entelis, E.M.Berliner, - M.:Mashinostroenie, 1986-352 p.

78. L.V.Khudobin, E.G.Berdichevsky. Technique of coolant application in metalworking. - M.: Mashchinostroenie, 1977. - 189 c.

79. A.K.Maskaev, I.L.Brovin, K.N.Kabilinsky. Areas of application of technological means in metal cutting processing - Kiev: Mashinostroenie, 1980 - 232 p.

80. Ya.I.Shkarapata, V.A.Soshko, V.G.Kriegel. Chemical composition of metal surface after mechanical processing // FCHMM. - 1990. - №4 - c.86-89

81. E.A.Stanchuk, A.I.Soshko. Increase of cutting tools durability by diffusion saturation during operation // Proc. of Nikolaev Shipbuilding Institute. Nikolaev Shipbuilding Institute. - 1981. - Vol. 174 - 68 c.

82. Y.I.Shulga, I.V.Kumpanenko , S.T.Entelis. Distribution of elements near the cutting edge of a drill // Surface. Physics. Chemistry. Mechanics. - 1982 - №3. - 135 c.

83. Gubanov A.K. Strength of Metals. FTT, 1964, vol. 6, pp. 1023-1073.

84. Bailey I.E.Electron Microscopy, v.9, Academy Press, L., 1962.

85. Stein D.L., Low I.R., Appl. Phys., 31, 362, 1960

86. Yong F.W., I.Appl.Phys., 32, 1815, 1962.

87. V.I. Mikheev, Hydrides of transition metals, M., ANSSR, 1960, p. 217.

88. K.R. Kapp. Uspekhi physics of metals. Vol. II, Metallurgizdat, 1958, p. 177.

89. R.A. Andrievsky. Ya.S. Umansky, Phases of implementation. Moscow: Nauka, 1977. 240 c.

90. P.V. Geld, R.A. Ryabov. Hydrogen in metals and alloys. M.: Metallurgy, 1974. 272

91. S.Z. Buckstein. Diffusion and structure of metals. M.: Metallurgy, 1973, 206 p.

92. Hydrogen in metals. Basic properties / Edited by G. Alefeld and I. Felkl. M.: Mir, 1981, 475 p.

93. B.S. Bockstein. Diffusion in metals. M.: Metallurgy. 1978, 248 c.

94. F.de Kazinczy. Jernkontorest annaler, 1955, p. 885

95. V.I. Tkachev. R.I. Kripyachev, A.K. Litvin. Friction and wear. FCHMM, 1966, NO. 3

96. H.H.Grimes, Acta metallurgica, 1959, 7, no. 12, 872

97. Y.N. Archakov. Y.I. Zvezdin, FCHMM. 1969, №5

98. M.A.Morris, M.Bowker, D.A.King, Chem. Kinetica, 1984, v.119

99. R.I. Kripiakevich, B.F. Kochmar, V.M. Sidorenko. Hydrogen in metals. FCHMM, 1970, No.5, p.95.

100. P.A. Mohr. Catalysis and inhibition of chemical reactions. M., "Mir". 1966. c. 257

101. Hydrogen in metals. Basic properties / Edited by G. Alefeld and I. Felkl. M.:Mir, ed. 2. 1992, 499 c.

102. H.H.Schumann Metallurgie und Gissereitechnik, 1933, B 4, S. 123

103. C.I.Smithells, Proc. Roy. Soc. London, A, 1935, v. 152, NA 877, p. 706

104. A.S.Tetelman, W.D.Robertson trans. AIME, 1962, v.224, No.4, p.775.

105. A.I.Soshko. Polymers in technological processes of metal working. - Kiev. - 1977, c. 115.

106. M.M.Shved, I.S.Slabkovsky, V.S.Senyushko - FCHMM: - No. 3, 1970 - p. 112

107. B.A.Shmelev. Methods of determination and research of the state of gases in metals. - M.: - 1968. - c. 48

108. M.A.Morris, M.Bowkes. - Chem. Kinetics. - 1984. - v. 19

109. P. Ashmar. Catalyzes inhibition of chemical reactions. - M.: Mir, - 1966 - p. 75

110. V.C.Venkatesh Furthor contribution to study of white lager on HSS tools. "Ann.CJRP" , - 1968, - № 2, p.15

111. E.A.Stanchuk. Processing of heat-resistant materials for ship turbine construction. Collection: "Shipbuilding and Marine Structures", vol. 10, Kharkov: - 1968, p. 119.

112. V.M.Zinchenko, V.V.Kuznetsov. To a question about a role of hydrogen at chemical heat treatment. - M.: Metallology and heat treatment of metals. - №2, - 1983. - c.59

Printed by Books on Demand GmbH, Norderstedt / Germany